BASIC MICROSCOPE TECHNIQUES

by

Philip Perlman

Director of Laboratory, Bureau of Supplies
Board of Education, New York City

CHEMICAL PUBLISHING COMPANY, INC.

New York   **1971**

Basic Microscope Techniques

ISBN: 978-0-8206-0007-9

Chemical Publishing Company:
www.chemical-publishing.com
www.chemicalpublishing.net

First edition:
Chemical Publishing Company, Inc. - New York 1971
Second Impression:
Chemical Publishing Company, Inc. - 2011

Printed in the United States of America

PREFACE

Basic Microscope Techniques endeavors to bring to the microscopist the modern basic microscope procedures he will encounter in the correct use of the microscope. The book was written to encompass students of the high school and college levels.

Theoretical concepts of each procedure are elucidated, and simultaneously, their practical aspects are fully covered. The worker thus obtains a comprehensive and clear understanding of what he is doing and consequently ensures the best use of the microscope.

The author has endeavored, throughout the manuscript, to simplify all details and make all instructions clear and foolproof so that the book could serve as a guide to microscopical procedures with all steps in a method fully delineated.

The parts of the compound microscope are described and its use is initially explained. Gradually, the employment of other specially designed microscopes and the understanding of advanced techniques with the optical microscope is developed. The worker becomes aware that the microscope is a very valuable tool in the various fields of science and industry and can be employed in the identification of inorganic and organic substances and additionally, for analytical procedures which ordinarily would be impossible or very time consuming.

A chapter on photomicrography has been included so that the worker may record, in an optimum manner, any desired microscopic specimens. It is believed that a great deal of satisfaction and enjoyment will be felt by the microscopist when obtaining his first photomicrographs.

A valuable portion of the book is the appendix. This lists the necessary reagents, their preparation and optimum use in the different phases of optical microscopy. A choice of fixatives, stains, etching reagents, photographic formulae, etc., is given to the worker from which he may choose a preparation that will give the microscopic specimen the best appearance.

I wish to thank the many optical and scientific supply companies for permission to include their many excellent reproductions.

CONTENTS

CHAPTER 1

The Compound Microscope

THOSE ENGAGED IN MICROSCOPY should be familiar with the basic construction of the compound microscope and have adequate knowledge of the basic principles of its use.

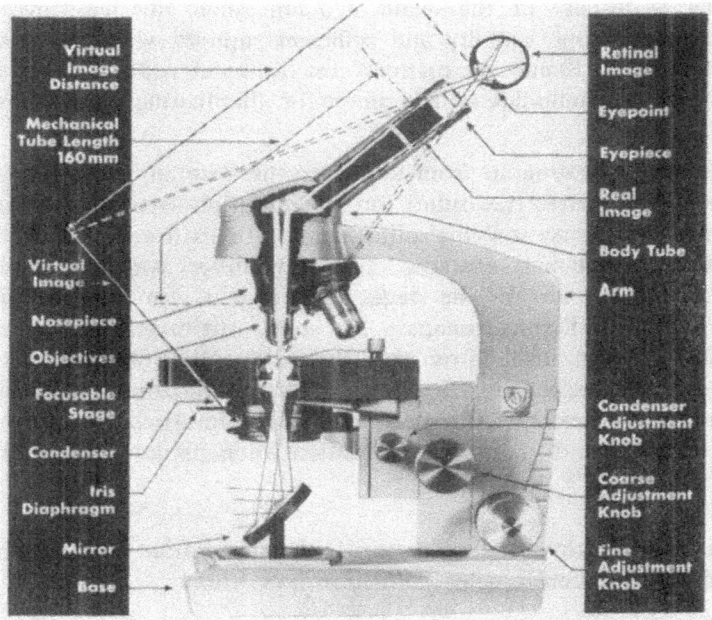

Virtual Image Distance

Mechanical Tube Length 160 mm

Virtual Image

Nosepiece

Objectives

Focusable Stage

Condenser

Iris Diaphragm

Mirror

Base

Retinal Image

Eyepoint

Eyepiece

Real Image

Body Tube

Arm

Condenser Adjustment Knob

Coarse Adjustment Knob

Fine Adjustment Knob

Fig. 1. Compound microscope.

The basic parts of the conventional compound microscope can be divided into three groups—the frame of the microscope with its mechanical parts, the magnification system, and the illumination system (see Fig. 1). The function of each group will be described so that the microscope can be used most efficiently and adapted to meet any special microscopic problem that may arise.

FRAME

The function of the frame of the microscope is to support the various optical and mechanical parts of the instrument so that it is durable and performs efficiently. The microscope rests on a heavy base which is usually in the shape of a horseshoe; this base gives the entire microscope stability and sufficient support when placed on a level surface. Some microscopes are equipped with a closed base that may accommodate a light source for illuminating the specimen.

An arm extending up from and over the base supports the body tube, which carries the optics (or magnification system), and incorporates the coarse and fine adjustments. In its lower part, the arm carries the stage and substage. The arm curves away to allow an unobstructed view of the stage. The arm is also used for lifting and carrying the microscope. On some instruments, there is an inclination joint in the lower arm that allows the microscope to be set at any angle from the vertical to the horizontal to facilitate comfortable viewing of the specimen. This joint also serves to incline the microscope to a horizontal position when photomicrographs are to be taken.

In the more advanced microscopes, the inclination joint has been eliminated and replaced by an inclined body tube.

Stage

The stage, carried by the arm, is a solid platform on which the specimen slide is placed for observation. In the center of the stage, there is a hole precisely centered on the optical axis of the instrument that allows light to be transmitted through the transparent specimen and through the center of all the optical parts.

Two clips on the stage hold the specimen slide in position.   A separate mechanical stage (see Fig. 2) can be attached to the permanent stage.   This mechanical stage holds the slide firmly and can be moved by a screw device that includes dual knobs; these enable the operator to move the glass slide slowly, smoothly, and accurately in two directions.   Such slow, steady movement of the specimen is essential when an object is being viewed under high power, since any rapid movement of the slide may cause the object to disappear from view.

Fig. 2.   Mechanical stage.

In many instances, the mechanical stage is built into the instrument. This useful accessory is provided with vernier scales graduated in millimeters, so that the movement of the specimen can be determined within 0.1 mm.   A graduated mechanical stage is very helpful when a specific portion of a slide is to be viewed again later.   This portion of the field can be determined by recording the readings on the scales.

Some modern microscopes have a "glide" stage, in which the entire upper stage slides on a film of grease and can be moved easily and precisely in any direction by a slight push of the fingers.

Body Tube

The body tube of the microscope serves to pass the light and the image formed by the objective to the eyepiece. A revolving nosepiece, designed to carry two or more objectives, is attached to the lower end of the body tube, and an eyepiece, which provides additional magnification and is used for viewing the specimen, is inserted in the upper end of the body or optical tube.

The length of the body tube on American-designed microscopes is fixed at 160 mm, measured from the top of the body tube, where the eyepiece is inserted, to the upper part of the objective attached to the nosepiece. Many foreign concerns have standardized their microscopes to a tube length of 170 mm. Other available microscopes have adjustable body tubes, the length of which can be varied.

Objectives are designed and standardized for use with a tube of specific length. If the tube length is wrong for the manufactured objective, the image may be poor in definition.

The objective is designed to be used with a cover glass of specific thickness (approximately 0.17 to 0.18 mm); any variation in thickness of the cover glass will cause a variation in the optimum tube length necessary for use with a particular objective. If the cover glass is thicker than required for the objective, it will then be necessary to shorten the tube in order to obtain a good image; if the cover glass is too thin, a longer tube will be required. It is therefore essential that the thickness of the cover glass be kept within the tolerance designed for the specific objective so as to derive the optimum value from a fixed 160 mm tube length.

COARSE AND FINE ADJUSTMENTS

The specimen slide is usually brought to a focus by moving the body tube of the microscope up or down. In modern microscopes, the focusing is done by moving the stage up or down while the body tube is kept immobile.

The coarse movement of the tube or stage is actuated by a device called the *coarse adjustment*, which brings the specimen slide into approximate focus. Exact focus of the object is then obtained with a device called the *fine adjustment*, which moves the body tube or stage very slowly.

The coarse and fine adjustment actuating knobs are usually situated on the arm of the microscope. The coarse adjustment may consist of a diagonally cut rack positioned on the body of the microscope and a pinion located on the optical tube or movable stage. When the coarse adjustment knob is rotated, the teeth of the pinion engage the teeth of the rack, thus moving the body tube or stage up or down.

The fine adjustment device may be a separate mechanism or, as in some microscopes, it may be concentric with the coarse adjustment and actuated by the same mechanism. The fine adjustment moves the body tube or stage very slowly so that the viewed object can be brought into very sharp focus.

In modern microscopes the fine and coarse adjustments are positioned low at the foot of the microscope and are thus more conveniently manipulated

On many microscopes the fine adjustment is graduated in units of 1 or 2 micrometers (1 micrometer$=1 \times 10^{-6}$ m), so that the distance of each complete rotation or fraction of a rotation can be easily determined. The graduated fine adjustment is used to measure the vertical dimension (that is, the thickness) of a specimen.

The modern microscope generally incorporates as a safety feature a mechanical stop, which prevents the objective lens from being unintentionally pushed into a specimen slide and possibly damaging both objective and slide.

Mirror

The mirror is located under the substage. A specimen cannot be seen under the microscope unless it is self-luminous or illuminated by an external source of light. The mirror therefore serves to reflect the external light to illuminate the object uniformly. The mirror is generally mounted on a fork that can be inserted into the pillar of the microscope. The mirror assembly allows the mirror to be moved in both the horizontal and the vertical plane or moved as a pendulum somewhat away from the optical axis of the microscope so as to reflect the light obliquely.

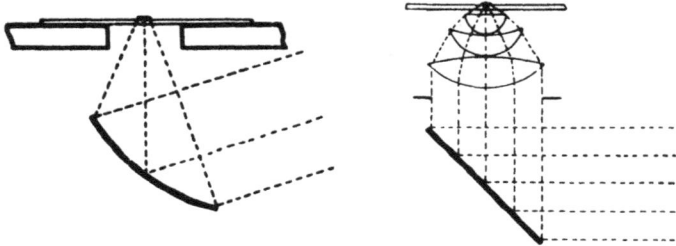

Fig. 3. Diagrammatic representations of the path of light reflected from a concave mirror (left) *and from a plane mirror with a substage condenser* (right).

The mirror may be two-sided, that is, with two reflecting surfaces. One side is a plane, or flat, mirror, the other a concave mirror. The latter, capable of gathering more light, is used when the microscope is not equipped with a substage condenser. The plane mirror is used when a substage condenser is part of the microscope (see Fig. 3, *right*). The condenser is more efficient in gathering and focusing the light to illuminate the object evenly. If the concave side of the mirror is used in conjunction with the condenser for high-power work, the mirror may produce a distorted image of the light source and prevent the condenser from focusing the light uniformly on the specimen. When the specimen is viewed under low magnification, however, the concave mirror gives reasonably good results when used in conjunction with a substage condenser.

Condenser

For best results, the compound microscope should be equipped with a substage condenser carried on a focusable mount. The focusable mount is necessary since not all microscopic slides are of equal thickness and the condenser must be moved up or down to focus the light for optimum uniform illumination.

The condenser is necessary especially for higher magnifications since without it the resolution of the image may be decreased. The resolving power of an objective without a condenser is reduced to approximately one-half. The condenser is simply a lens or series of lenses that gather light from the mirror or illuminating source and concentrate it on the specimen slide. The focal distance is about 1.25 mm above the hole, or aperture, of the stage.

Fig. 4. Abbe condenser.

The Abbe uncorrected condenser (see Fig. 4) is the one most generally used in routine microscopy. It usually consists of two lenses mounted in a single case. The upper one can be unscrewed. When properly used, this condenser can have a numerical. aperture of 1.30 if immersed in oil or other liquids. Three-lens Abbe condensers function at a maximum numerical aperture of 1.40. The Abbe condenser, which is not corrected for spherical or chromatic aberration, is adequate for most work where color correction of the transmitted light is not required or the exact image of the lamp source is not essential.

The upper lens of the Abbe condenser can be removed. This permits using low-power magnification with the lower lens of the condenser, which has a longer focal distance when used alone.

Other types of condensers have as many as six elements and are corrected to the same degree as objectives. The corrected condensers give a good undistorted image of the light source with less glare.

The aplanatic condenser, which is corrected for spherical but not chromatic aberration, is desirable for photomicrographs; it performs well with light of one color.

The achromatic condenser which is corrected for chromatic and usually spherical aberration, gives the highest performance, since its precision focusing eliminates glare around the specimen.

A good condenser, therefore, is one that transmits the correct illumination to the viewed object and gives better contrast by eliminating stray light rays, which would cause glare. Stray light can be defined as light that is transmitted through the optical system but does not take part in the illumination and formation of the image.

Iris Diaphragm

The substage iris diaphragm is positioned beneath the condenser. The iris which is operated by means of a small lever projecting from the diaphragm and extending beyond the circumference of the condenser, is employed to regulate the transmitted light to a sufficient and necessary amount.

If the iris diaphragm is opened too fully, the intense illumination will cause glare and undesired light reflections within the microscope. Halos may also occur at the edges of the field of view damaging the resolution of the specimen image. If, on the other hand, the iris diaphragm is closed down too much, the numerical aperture of the objective is simultaneously reduced. This not only decreases resolution of the specimen, but also causes fringes to appear around some details of the specimen because of the interference of the light rays.

The correct aperture of the iris diaphragm, therefore, permits control of the light beam so as to attain the best contrast of the specimen and the maximum resolving power of the objective.

Once the correct opening of the iris diaphragm has been determined for the objective (see Chapter 2), it should not be altered. If the intensity of light is too strong, the light can be reduced by inserting a neutral-density optical filter or colored filter in the filter holder beneath the diaphragm.

OBJECTIVES

The objectives (see Fig. 5), which are used for initial magnification of the specimen, can be considered the heart of the microscope, since the performance of the instrument depends on the quality and design of these lenses. These lenses are called *objectives* because they are positioned above the object to be magnified, and their chief function is to produce the primary magnified image of a specimen.

The revolving nosepiece, which is attached to the lower end of the body tube, carries two, three, or four detachable objectives of different powers; they have marked or engraved upon them, their magnifying power, rated numerical aperture (N.A.), equivalent focal length, and sometimes whether they are achromatic or nonachromatic. An example of the legend inscribed on the barrel of an objective is: Achromatic-16 mm, 10X, N.A. 0.25. The meaning of these terms is more fully discussed below.

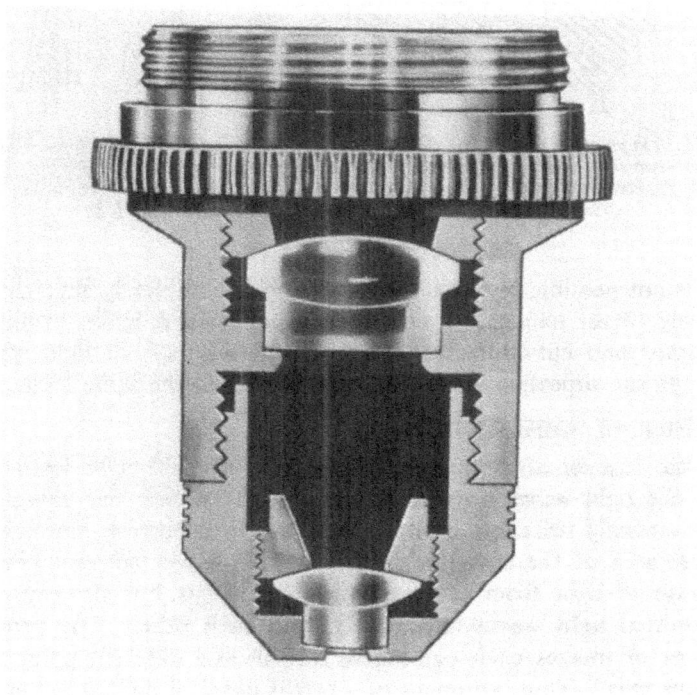

Fig. 5. Cross section of a typical objective.

Types of Aberration

The construction of an objective lens is complicated. Elements of different curvatures (see Fig. 6) and made of different types of glass may be used, since the objective must be as free as possible from such defects or aberrations as chromatic, spherical, astigmatic, and curvature of field. Objectives chosen for the microscope must therefore be corrected, partially or totally, for these aberrations so that the viewed object will be sharp and clear.

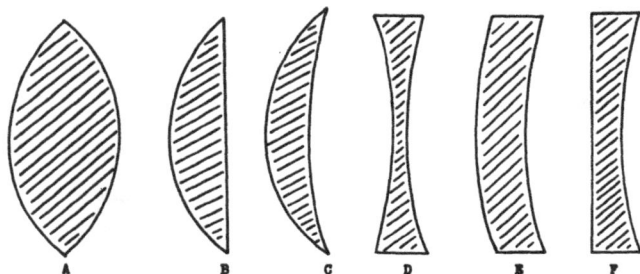

Fig. 6. Types of lenses : A) *double convex —positive, converging lens ;* B) *plano-convex—positive, converging lens ;* C) *concavo-convex—positive, converging lens;* D) *double concave—negative, diverging lens ;* E) *convexo-concave—negative, diverging lens ;* F) *plano-concave—negative, diverging lens.*

It is not possible to correct all aberrations completely and simultaneously. Most defects are corrected by combining lenses of different materials and curvatures so that the different rays of light passing through the objective will be brought almost to the same focus.

SPHERICAL ABERRATION

In the case of an optical double convex lens with spherical aberration, the light waves passing through the lens near the margin are more strongly refracted or bent than the rays passing through the central area of the lens (see. Fig. 7). This defect not only prevents the rays of light from coming to a sharp focus, but also causes the transmitted light waves to interfere with each other. This produces a series of images each coming to a focus at a different point along the lens axis. Thus, an image of a bright point of light is not seen as such, but rather will appear as a bright disc surrounded by dark and

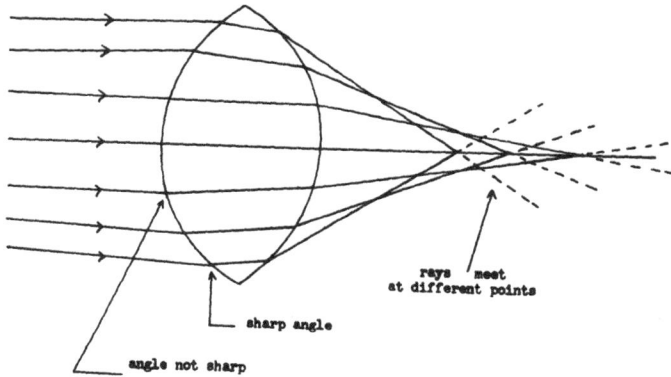

rays meet
at different points

sharp angle

angle not sharp

Fig. 7. Diagrammatic representation of spherical aberration in a double lens.

light rings forming a diffraction pattern.   Such an image is fuzzy and distorted and lacks sharpness.

A lens corrected for spherical aberration will concentrate the major portion of the light so that very little is lost in light and dark halos surrounding the image and will therefore produce a better-defined and crisper image of the object.   Complete correction for spherical aberration is not possible; it can be achieved only for one selected portion of the spectrum.   Achromatic objectives are corrected spherically for one color (generally yellow-green, in the midportion of the spectrum); it is therefore advantageous to use a green filter or green monochromatic light to obtain a sharper image.

CURVATURE OF FIELD

This defect in the objective will show the formed image of a flat specimen not as flat, but rather as curved. This curvature occurs because the image is not in sharp focus throughout the field of view. An objective with this defect will generally show the image in focus and better defined in the center of the field of view, while the image near the margin of the field will be out of focus. On the other hand, if the object is brought into sharp focus at the margin of the field of view, the image will then appear blurred and out of focus at the center of the field.   This defect is intensified with high-power objectives and great magnifications.

Objectives are usually made so that a reasonably flat field, adequate for routine work, is obtained over a large portion of the field in view. For photomicrographic work, however, where flat fields are essential for good rendition of the image, special flat-field objectives may be employed. Special eyepieces are also used to compensate for curvature of the field due to defects in the objective and to give an undistorted flat field. It must be emphasized that flat fields can be achieved only at the expense of resolution.

ASTIGMATISM

This defect in the lens system causes the image of a point to be spread or drawn out as a line in one or two directions (see Fig. 8, *left*). This defect can be demonstrated in a lens that has it by manipulating the fine adjustment. Two separate line images will be seen at right

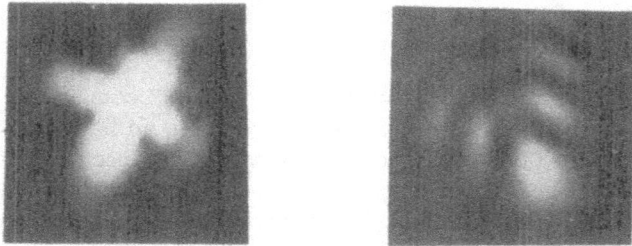

Fig. 8. Photomicrographs of the image of a point object produced by astigmatism (left) *and by coma* (right).

angles to each other. This defect is generally caused by irregular curvature in one or more of the lenses that make up the objective. Astigmatism may occur to a slight extent even in adequately corrected objectives, around the margins of the field of view but not in the midportions. In many instances, with fine adjustment, it is possible to misinterpret the shape of the bodies located at the margin of the field where round bodies will appear as rods.

COMA

This defect causes the objective lens to give different magnifications at different concentric zones of its lens surface. As a result, points of the specimen image appear comet-shaped and the specimen image is impaired (see Fig. 8, *right*). This defect can be easily corrected and controlled by the manufacturer by proper design of the objective lens surfaces.

LATERAL COLOR

A chromatic difference in magnification occurs when white light, normally composed of many colors, is used to illuminate the specimen. This defect results in the observed light of one color being in greater magnification than the light of another color, so that the off-axis image of a point in the specimen is spread out in color or as a tiny spectrum. Lateral color is more prevalent in high-power objectives and is difficult to correct, since in these objectives correction for more important defects, such as chromatic aberration, interferes with the simultaneous correction of lateral color.

This defect can nevertheless be easily corrected by using special compensating eyepieces that have been overcorrected for lateral color and are especially used with high-power objectives that have been overcorrected chromatically.

CHROMATIC ABERRATION

ACHROMATIC OBJECTIVES

The achromatic objectives which are most often used for routine microscope work are also adequate for many research procedures. The achromatic objectives are corrected mainly for distortion caused by color, that is, chromatic aberration. A simple uncorrected lens does not bring white light, which is normally composed of many color components of different wavelengths, into sharp, clear focus. Thus, white light, as it passes through a lens, is broken up into the colors of the spectrum. The colors range from violet, which has a shorter wavelength and focal distance, to red, which has a longer wavelength and focal distance. The violets are refracted, or bent, the most; the longer wavelength reds are refracted the least.

Under these circumstances, each color will come to a different focus because of the differential refraction of the components of white light. This phenomenon of chromatic aberration will cause unpleasant colored halos and images to appear around the magnified specimen. As a result, the viewed specimen will appear fuzzy and indistinct. This occurrence is quite objectionable, since the resolution of the specimen is impaired and some of the fine details cannot be seen.

Achromatic objectives are generally corrected chromatically for two colors, once in the red range, characterized by longer wavelengths, and again in the green spectral range, characterized by shorter wavelengths.

A corrected objective is produced by combining different types of glass, with different refractive indices, which bring the rays of light with the highest wavelength (red) and those with the shortest wave-

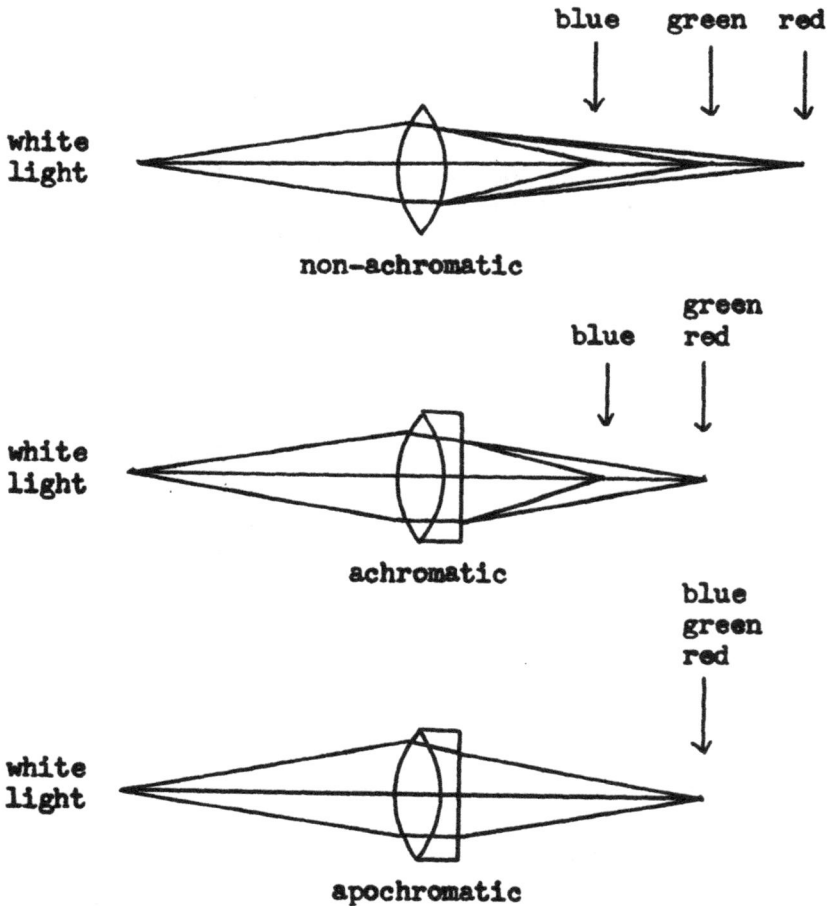

Fig. 9. Diagrammatic representation of the correction of chromatic aberration.

length (green) to a common focus (see Fig. 9). Under these circumstances, there is a good compromise obtained in focusing the intermediate colors in the visible spectrum.

The achromats are also corrected spherically for one color, yellow-green light. Achromatic objectives, therefore, are generally designed to give best performance with green illumination.

APOCHROMATIC OBJECTIVES

Apochromatic objectives are designed to give a higher degree of chromatic correction; they are also more fully corrected for spherical aberration. They are used especially in photomicrography, where the highest resolution of detail is most essential. The apochromats are corrected achromatically for three colors—the red, green, and blue spectral regions—and spherically for two colors—the green and blue spectral regions. The apochromats are made by combining several fluorite lenses with special barium flint-glass lenses. Fluorite is a calcium fluoride mineral of low refractive index especially adapted for optical lenses.

The apochromatic objectives, with a higher numerical aperture, are superior to achromatic objectives. Apochromats produce an image with greater contrast and brightness even when stained specimens are being viewed. Apochromatic objectives, however, do have an inherent design defect—"chromatic difference of magnification," or lateral color. When a specimen is viewed with an apochromatic objective, the blue and red formed images are of different sizes and as a result cause a lack of sharpness in the viewed specimen. To correct this defect, special compensating eyepieces are used in conjunction with these objectives so that the specimen image will be almost entirely free from color images and simultaneously produce a flatter field.

Low-power compensating eyepieces are the Huygenian type; eyepieces with a magnification of 15X or greater are the Ramsden type.

SEMI-APOCHROMATIC OBJECTIVES

Semi-apochromatic objectives are fabricated with fewer fluorite lenses than regular apochromatic lenses. They are not so fully corrected optically as the apochromats, so that a compromise in performance between the achromats and the apochromats is achieved. The semi-apochromatic objectives are rather popular because of their comparative low cost, and they are almost equal in performance to the apochromats when green color filters are to be used.

Working Distance

The working distance of an objective is defined as the clearance distance between the lowest part of the objective and the top of the cover glass when the specimen is in sharp focus. (The cover glass is included in this definition because the objective is generally corrected for use with a cover glass.) As a general rule, the greater the initial magnifying power of the objective, the shorter the working distance. However, special objectives designed to give extra working distances are also commercially available.

Great care must be observed when viewing a specimen with a high-power objective, since there will be little clearance between the specimen and the objective lens. There is the danger of ramming the objective into the specimen slide, a mishap that may break the slide and damage the objective lens.

Numerical Aperture

The numerical aperture (N.A.), or light-gathering capability of a lens (first expressed mathematically by Abbe), is of fundamental importance in microscopy. The ability of the objective to separate the fine details of an object under magnification is dependent on its numerical aperture. The greater the numerical aperture of an objective, the finer the detail observable in the viewed specimen (see Fig. 10). Magnification by itself does not necessarily increase the resolution of a specimen. It is possible to magnify an object to an extent where no further detail is revealed and the sharpness of the image simultaneously deteriorates. This type of magnification is called *empty magnification*. The upper limit of the true usable magnification of a specimen is determined by the wavelengths of the light employed to illuminate the specimen and the numerical aperture of the lens system.

The numerical aperature can be expressed as a numerical quantity that characterizes the ability of a lens to allow a greater cone of useful light, illuminating the specimen, to pass through the lens or objective. The greater the amount of light, or, technically, the greater the angle of illumination coming from the specimen and utilized by the objective, the greater the ability of the objective to resolve details in the specimen.

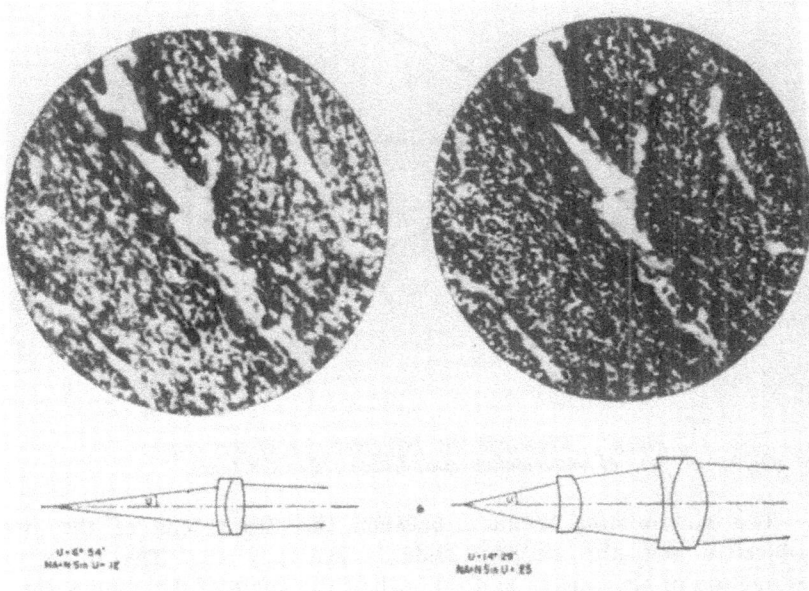

Fig. 10. Photomicrographs and diagrams illustrating the dependence of resolving power on the numerical aperture (N.A.). The N.A. of the left-hand objective is 0.12; of the right-hand objective, 0.25.

The quantity of numerical aperture can be expressed mathematically as $N.A. = n \sin \mu$, where $N.A.$ is the numerical aperture, n is the refractive index of the medium between the cover glass on the specimen slide and the front lens of the objective, and μ is one half of the angle of the cone of light passing through the lenses of the objective, also expressed as half the angle of illumination.

Figure 11 illustrates the concept of numerical aperture. A source of light emanates from an object, point A, and passes through the lens, BD. The symbol μ indicates the half angle of the cone of light that enters the lens, AC is the focal length of the lens, and BD is the diameter of the lens.

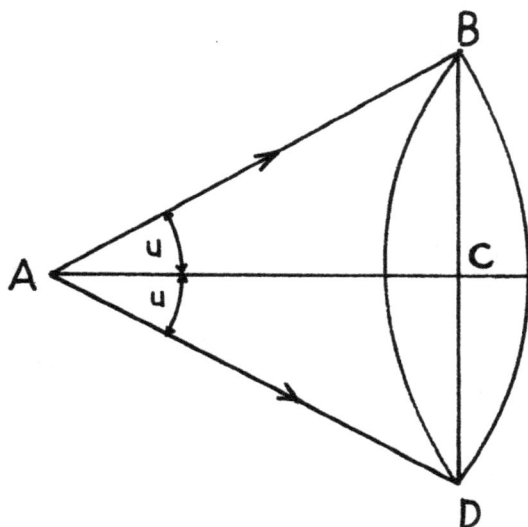

Fig. 11.   Diagrammatic representation of the concept of numerical aperture (discussed in the text).

The surrounding medium between the front lens of the dry objective and the specimen slide is generally air. The index of refraction of air is taken as 1.00; μ, half the angle of the cone of light rays entering the objective lens, can approach, at most, 90°. Mathematically, the sine of 90° is 1. It is apparent from the preceding equation that the numerical aperture of a dry objective, when the medium between the objective and the slide is air, cannot exceed the numerical value of 1.

In order to increase the numerical aperture to a value greater than 1.00, it is necessary to insert between the specimen slide and the front lens of the objective a few drops of a medium that has a higher index of refraction than air—for example, a liquid such as water, which has an index of 1.33, or an immersion oil with an index of 1.55. The medium is placed on both the slide and the front lens of the objective, so as to immerse both in the liquid. Objectives designed especially to be used only in this manner are called *oil-immersion objectives*.

Resolving Power

The resolving power of a microscope can be defined as the capability of the microscope to distinguish fine detail in a specimen or—more precisely—to distinguish two adjoining points of a specimen (minimum separation) as distinct and separate entities so that they can be seen and recognized as such and not as one fuzzy point.

If light coming through a very small pinhole in an aluminum mirror is photomicrographed using a microscope and camera, it will be found that the recorded image of the light is not a point, but is actually a small circular spot of light surrounded by darker rings of light. This phenomenon is due to the diffraction of the light as it passes through the pinhole. This pattern of light, called the *Airy disc*

Fig. 12. Photomicrograph of an Airy disc.

(see Fig. 12), was named after the English astronomer Sir George Airy, who showed that by measuring the radius of the first dark ring, he could determine the resolving power of a lens. The Airy disc has been satisfactorily used to study different lens aberrations or defects.

The resolution of an objective, however, also depends on the wavelength of the light used to illuminate the specimen and on the numerical aperture of the objective. The resolution is expressed mathematically as $R = 0.61\lambda/N.A.$, where R is the limit of resolution (minimum distance between two points of an object that can be distinguished as such), λ is the wavelength of the light used, and $N.A.$ is the numerical aperture of the lens or objective.

From the equation, it can be seen that the smaller the wavelength of light used and the greater the numerical aperture of the optical lens system, the smaller the minimum distance necessary for two adjacent points of an object to be separated and seen as two points.

For example, assume that two dots are observed with one objective as two entities when they are 0.0001 mm apart. This objective has a greater resolving power than another objective, which observes the dots as two entities only when they are at least 0.0002 mm apart. It is evident, therefore, that the resolving power of the microscope will determine how much fine detail can be seen in the specimen.

The resolving power of a microscope can be increased by decreasing the wavelength of the light used to illuminate the specimen. Green light and, to a lesser extent, light in the blue-violet region are often used to improve resolution. Light of even shorter wavelengths, such as those in the ultraviolet regions, may also be used, but special optical lenses made of quartz must be used to permit transmission of the ultraviolet radiation. Since, moreover, the eye is not sensitive to ultraviolet light, the image must be recorded photographically.

The useful characteristics of objectives are listed in Table 1.

TABLE 1. AVERAGE CHARACTERISTICS OF ACHROMATIC, SEMI-APOCHROMATIC, AND APOCHROMATIC OBJECTIVES WITH A 10X EYEPIECE

Type	Initial Magnification	Focal Length (mm)	Numerical Aperture	Working Distance (mm)	Diameter of Field (mm)	Depth of Focus (micrometers)
Achromatic						
Dry	10	16.0	0.25	4.5	1.5	to 10
Dry	40	4.0	0.65	0.5	0.35	1 to 2
Oil-Immersion	95	1.8	1.25	0.13	0.15	0.5
Semi-apochromatic						
Dry	44	4.0	0.85	0.52	0.29	1 to 2
Oil-Immersion	97	1.8	1.30	0.10	0.14	0.5
Apochromatic						
Dry	10	16.0	0.30	5.2	1.29	to 10
Dry	44	4.0	0.95	0.20	0.29	1 to 2
Oil-Immersion	90	2.0	1.30	0.10	0.14	0.5
Oil-Immersion	90	2.0	1.40	0.05	0.14	0.5

EYEPIECE (OCULAR)

The eyepiece, or ocular, which is located in the upper part of the optical tube, serves to increase the magnification of the primary image formed by the objective and at the same time to render this formed image visible. Eyepieces are classified as positive or negative. The negative eyepiece forms a real image between its lens elements; the positive forms a real image below the lens elements of the eyepiece. Magnification powers of the eyepieces most commonly used for routine microscopy are 5X, 10X, and 15X, with the 10X being the more widely used.

Huygenian Eyepiece

The most common and most versatile type of eyepiece used for general and routine microscopy is the Huygenian eyepiece (see Fig. 13), a negative type consisting of two plano-convex lenses with the convex sides of both lenses directly facing the object to be viewed. The lenses are separated about midway in their housing by a circular diaphragm, called the *field diaphragm*, inserted between the two lenses and soldered to the barrel of the eyepiece. The field diaphragm is a circular piece of thin metal with a precise aperture to limit the size of the field. It is positioned at the focal point of the upper lens, which is called the *eye lens*.

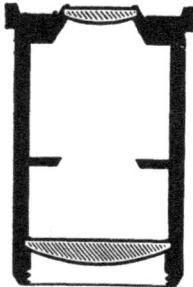

Fig. 13. *Huygenian eyepiece.*

The lower plano-convex lens, called the *field-lens*, is used to gather the rays of light of the magnified image transmitted by the objective and to focus them on or near the field diaphragm. The upper, or eye, lens magnifies this focused image, which can then be viewed by the eye.

Thus, the field diaphragm not only serves to limit the size of the field of view, but also cuts off any stray rays that would impair the sharpness of the image. The size of the aperture of the field diaphragm can be manipulated to give a larger field of view. The larger the aperture of the field diaphragm, the greater the field of view. The resolution of the specimen image, however, will suffer with any substantial increase in this aperture. The optimum size is generally fixed by the manufacturer. Wide-field eyepieces giving a wider view of the specimen are commercially available; they are primarily used to search a specimen slide for specific features or to observe large specimens stereoscopically.

The diaphragm in the Huygenian eyepiece serves also as a platform on which microscopic accessories such as counting or measuring scales can be placed. A hair can be cemented to the platform to serve as a pointer for teaching.

Hyperplane, periplane, and orthoscopic eyepieces are similar in design to the Huygenian eyepiece, but are corrected for chromatic aberration and lateral color and give a flatter field of view.

Ramsden (Kellner) Eyepiece

The Ramsden positive eyepiece is not as widely used as the Huygenian. It consists of two plano-convex lenses of similar focal length, with the convex surfaces facing each other. With this type of eyepiece, the primary image is formed below the field lens, thus forming a positive image. The diaphragm is therefore situated below the lower lens, that is, the field lens, so that a measuring scale can be placed on the diaphragm. The Ramsden eyepiece is most often employed for measuring and for purposes of comparison.

Compensating Eyepieces

Compensating eyepieces can be either negative or positive. They are special eyepieces designed to be used in conjunction with apochromatic objectives, especially high-power objectives; their function is to correct certain inherent defects of these fine objectives. The eyepiece compensates for any remaining color aberration still present in the objective due to the hemispherical shape of the front lens of the apochromats, the design of which produces different sizes of red and blue images that cause lack of sharpness (lateral color).

The compensating eyepiece consists of two convex lenses for the field lens and one plano-convex lens for the eye lens, with all the curved surfaces facing downward. This type of eyepiece gives a flatter field, has a greater initial magnification than the standard eyepieces, and is overcorrected for lateral color.

Photomicrographic or Projection Eyepieces (Amplifier Type)

Special negative eyepieces are used in photomicrography to correct the curvature-of-field defect, which is inherent in most objectives. Projectars, Ampliplans, Homals, and Ultraplans are a few of the amplifier type of eyepieces that are used strictly for photomicrographic work. They are not suited to visual work, since their exit pupils lie within the lens elements.

Binocular Eyepieces

The binocular microscope employs a matched pair of Huygenian eyepieces but retains the single objective of the standard monocular type. A system of prisms and mirrors, which is part of the binocular body tube, splits the incident beam of light coming from the objective and produces two identical images that can be viewed simultaneously by both eyes (see Fig. 14). The use of both eyes lessens fatigue and eyestrain on the part of the observer and also gives some illusion of depth. The eyepieces can be separated and adjusted to the interpupillary distance of the viewer. It must be observed that the intensity of illumination is lessened, since only half of the light is transmitted to each eye.

DEPTH OF FOCUS

When the specimen is brought into focus with the microscope, only a limited thickness of the object is in focus. The magnitude of this thickness called the *depth of focus* depends on the numerical aperture of the optical system and the magnification used. Generally speaking,

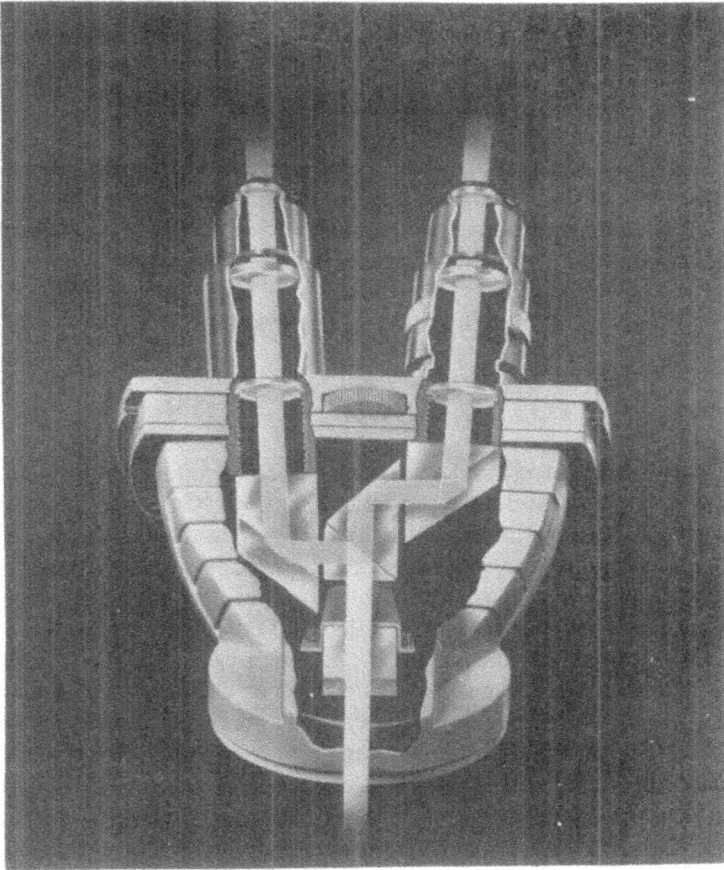

Fig. 14. Cutaway photograph showing the path of light through a binocular microscope.

the depth of focus decreases as the magnification and the numerical aperture increase and as the wavelength of the illumination decreases. When examining a specimen for such general characteristics as thickness, size, and arrangement of structures, objectives of low power, longer focal length, and with a smaller numerical aperture should therefore be used to give more satisfactory results, although the resolution of the specimen will be lessened. When very thin sections are to be examined for minute details, objectives of high numerical aperture should be used.

DIAMETER OF FIELD OF VIEW

The area of the field depends on both the eyepiece and the objective. The diameter of the field of view becomes smaller as the magnification of the chosen objective increases. For any specific objective, moreover, the field of view decreases as the magnifications of the chosen eyepiece increases. The effect of the objective on the field of view is, however, greater than that of the eyepiece.

TOTAL MAGNIFICATION

The major purpose of the microscope is to magnify the specimen sufficiently so that its details can be clearly seen. Generally speaking, the magnification power of the instrument is determined by comparing the size of the object as seen with the microscope with its size as seen with the naked eye. A person with normal eyesight, when observing an object or reading a paper, will hold the object or paper at a distance of 10 in. (254 mm) from the unaided eye. At this distance, the normal eye will focus sharply and observe the object or scan the printed paper with the least difficulty. The magnification power of a microscope is determined in a similar manner by measuring the size of the magnified image projected 10 in. from the eyepiece. The apparent size of the projected image is compared with the original dimensions of the magnified object and the proportion is generally expressed in number of diameters.

The total magnifying power of a microscope can be more easily determined by multiplying the initial magnification power of the objective by the additional magnification power of the eyepiece. For example, if the power of the eyepiece is 10X (called *ten power*) and that of the objective is 40X, then the total magnifying power of the microscope is 400X. This means that an object viewed under a microscope with a 40X objective and a 10X eyepiece will be enlarged to 400 times its original diameter. Different magnifications can be achieved by varying the power of the eyepiece.

The total maximum magnification of a specimen, however, should never exceed 1,000 times the numerical aperture of the objective. Similarly, for optimum results, the magnification should not be less than 250 times the numerical aperture of the objective selected. These empirical figures are workable for lenses that are free from most optical defects and for persons with good normal eyesight.   A more realistic figure generally used by microscopists holds that the maximum total magnification should not exceed 750 times the numerical aperture of the objective.

There are other factors that must be considered in determining the desirable magnification for a specific specimen.   When a specimen is observed under high magnifications, the field of view will be lessened, together with the working distance and the depth of focus.   Consequently, when high magnifications are employed, greater precision is required for the preparation of the specimen.

COATED OPTICS

The optic parts of modern microscopes (condenser, objectives, and eyepieces) are generally coated with a thin (0.14 μ) film of magnesium fluoride to prevent the loss of incident light. On the optical lens, this coating reduces the reflection and absorption of light and at the same time allows greater transmission of useful light through the microscope.

CHAPTER 2

Methods of Illuminating and Using the Microscope

PROPER ILLUMINATION is one of the most important factors in improving the quality and contrast of the image viewed through the microscope. Adequate, uniform illumination is essential, and the illumination system must be simple and easily adjusted. Optimum resolution of the specimen by the objective, so that the image appears crisp and detailed, is also basically dependent on proper illumination.

DAYLIGHT ILLUMINATION

For elementary or routine work, ordinary daylight, the light most readily available, is a good source of illumination provided its limitations are understood. The major objection to natural daylight is that it is difficult to control the intensity for optimum illuminating conditions. Daylight is a diffuse form of illumination. Nevertheless, there is the possibility that some glare may occur if the specimen is illuminated with direct sunlight. If the aperture or lens opening of the objective is properly illuminated, however, no difficulty should occur when this source of light is used.

The microscope is placed near a large window so that the diffused daylight from the sky or clouds is reflected from the microscope mirror onto the specimen. Direct sunlight or light coming from near the sun should never be used.

ARTIFICIAL ILLUMINATION

Daylight illumination is too inconsistent to be dependable for anything other than elementary or routine work. The microscope cannot be utilized at night or on cloudy days when the intensity of natural light is inadequate. Consequently, an artificial light source is preferred by most workers with the microscope, since the source of light can be more readily controlled and uniform illumination of the specimen can be achieved.

For routine work with the microscope, artificial illumination transmitted by a simple, inexpensive filament microscope lamp is sufficient. The intensity of light from these lamps can be easily controlled by varying the distance between the lamp and the microscope mirror. For research work, however, where special lighting problems are involved or where great fineness of detail is desired in a specimen for metallography, photomicrography, or microprojection, a more advanced type of illuminator is essential for best results. Some of the more common types of microscope illuminators are described below.

Microscope Illuminators

SIMPLE SUBSTAGE ILLUMINATOR

The simple substage illuminator (see Fig. 15), which may vary in design with different manufacturers, usually consists of a metal or plastic lamphouse with ventilating louvers. The lamphouse encloses a high-intensity filament lamp bulb and a reflecting surface. A heavy glass filter is attached to the front housing of the lamp. The filters may be of ground or frosted glass to diffuse the light or of blue Corning daylight glass to simulate natural daylight. The illuminator is provided with a built-in switch and connecting cord.

The lamp usually holds a 10- or 15-watt bulb with a reflector behind and a blue ground glass in front. In many microscopes the mirror has been supplanted by a built-in illuminator at the base of the microscope. Attachable illuminating sources, interchangeable with the mirror, are also available. The attached illuminator is advantageous in that the lamp can be permanently aligned and is easily operated.

Fig. 15. Simple substage illuminator.

ADVANCED SUBSTAGE ILLUMINATOR

The advanced substage illuminator (see Fig. 16) is convenient for both routine and advanced research work. The design generally embodies a ventilated die-cast metal housing, usually aluminum, mounted on a stable base with an adjustment for tilting the illuminator. The light source may be a tungsten multifilament, 120-volt lamp. The illuminator may be supplied with a transformer switch to regulate the voltage, which determines the intensity of the light. A variable-focus

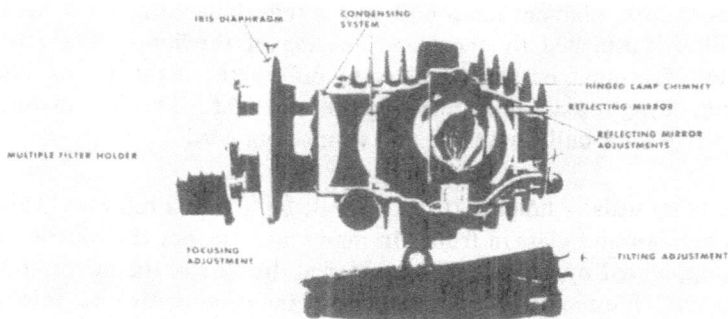

Fig. 16. Advanced substage illuminator.

parabolic condenser lens with a focusing adjustment knob is part of the illuminator.

The focusing knob can be rotated to refocus the light for high- or low-power objectives. The microscope illuminator may be preset for two positions. When preset in the forward position for low-power objectives, the condenser is moved away from the lamp source; in this position, the illumination is sufficient to fill the back lens of the objective completely, so that the field of view is completely and uniformly illuminated. For high magnifications, the illuminator is preset in the second position, with the condenser closer to the light source. As a result, the lamp source is magnified and, when transmitted to the substage microscope condenser, results in a smaller illuminated aperture of the lens. This gives more uniform illumination of the smaller fields of high-power objectives. An iris diaphragm, sometimes graduated, controls the amount of light reaching the specimen.

The lamphouse includes as standard equipment a holder and various types of round or square filters. These include a "daylight" filter, which supplies light closely approximating daylight; a set of neutral density filters, which are used to reduce the intensity of illumination (available in 5, 10, 20%, etc., transmission); and a heat-absorbing filter, which may be used instead of cooling cells when examining specimens that are sensitive to heat. Special illuminators are also available for particular problems as they arise. Some of them are described below.

CARBON ARC LAMP

The carbon arc illuminator (see Fig. 17) is often chosen when an extremely intense light source is necessary, as in high-power photomicrography. The arc lamp provides the white light required for identifying colors within a specimen. It consists of two carbon electrodes mounted at right angles and can be operated on either direct or alternating current. Direct current is usually preferred, since it gives a steadier light without flickering. Since constant distance between the electrodes must be maintained to ensure steady illumination, good arc illuminators have some means of automatically feeding the electrodes, such as an electric motor.

Fig. 17.   Carbon arc lamp.

POINTOLITE LAMP—SPOTLIGHT ILLUMINATOR

The spotlight illuminator (see Fig. 18), with a transformer for varying light intensity, provides an intense "spotlight" effect. It is used mostly for lighting small crevices, holes, or other irregularities in the specimen, or for pinpointing intense light on a specific area.

MERCURY-VAPOR LAMP

The mercury-vapor lamp (see Fig. 19) provides an intense source of monochromatic light in the green (546 nanometers), blue (436 nanometers), and ultraviolet (365 nanometers; 1 nanometer$=10\times$ 10^{-9} m), wavelength regions. These monochromatic colors are achieved by means of special filters that allow only the desired wavelengths of light to be transmitted.

SODIUM-VAPOR LAMP

The sodium-vapor lamp (see Fig. 20) emits yellow monochromatic light at the 589-nanometer wavelength region of the spectrum, commonly known as the D line. Actually, it consists of two very close lines as observed on a spectroscope (588.99 and 589.59 nanometers).

TUNGSTEN AND ZIRCONIUM ARC LAMPS

Tungsten and zirconium arc lamps give very intense light especially suitable for photomicrography. The former is generally used for black-and-white photography, the latter for photomicrography in color.

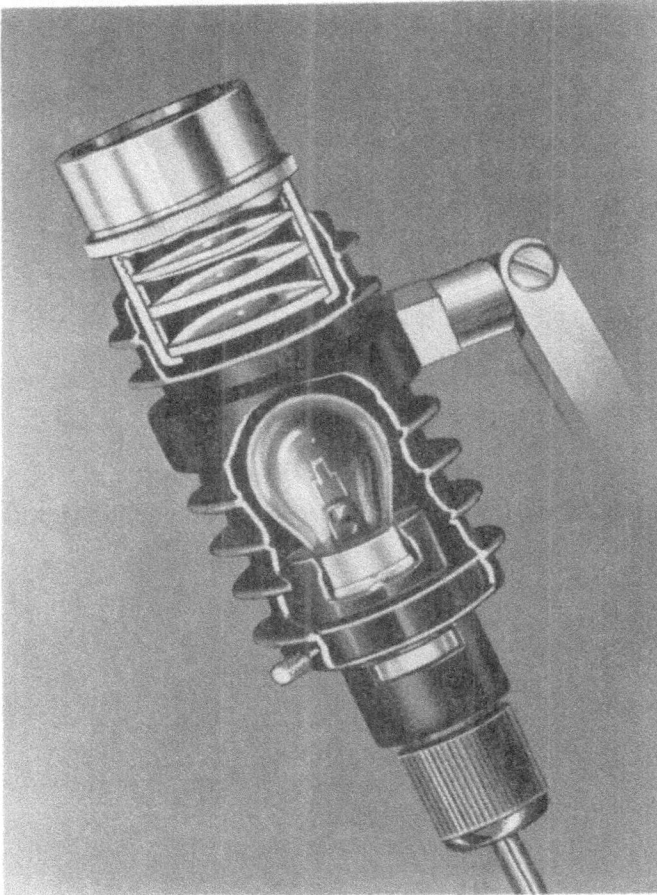

Fig. 18.   Spotlight illuminator.

FIBER OPTICS

Illumination of a microscope stage by "cold light" can be achieved with a flexible fiber-optic accessory by piping the cold light from a distant source to a biological specimen under study. This type of light is advantageous in that it eliminates the heat given off by an ordinary light source, which might affect or destroy living or dead specimens.

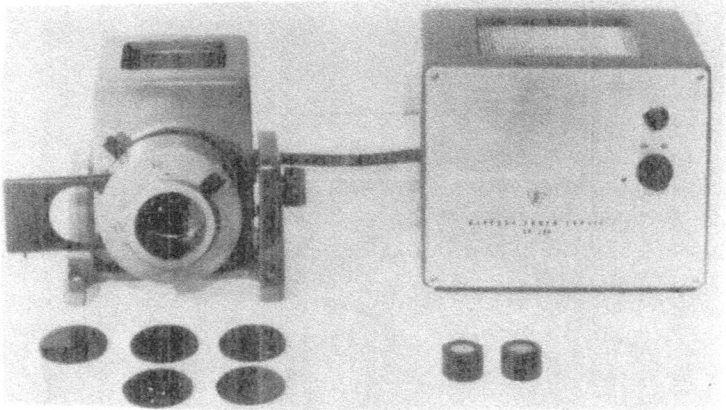

Fig. 19. Mercury-vapor lamp.

Bright-Field Illumination

Most routine work with the microscope is accomplished with bright-

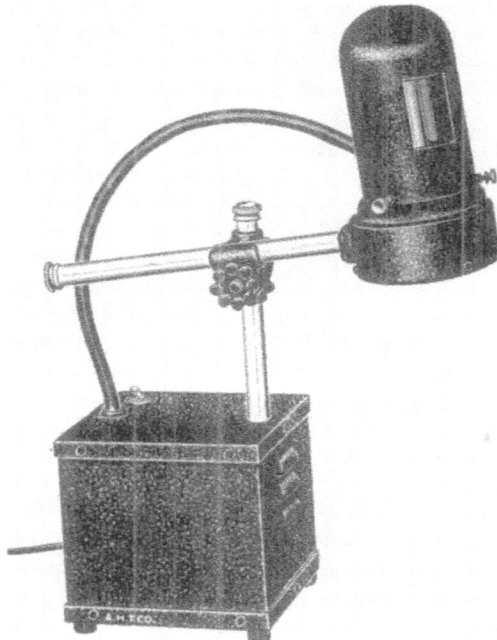

Fig. 20. Sodium-vapor lamp.

field illumination, in which the light is transmitted through the condenser lens and focused on the specimen. With bright-field illumination, the object specimen appears more or less dark against a lighter background. In most cases, good contrast between specimen and illuminated background is achieved, so that good resolution is obtained.   Bright-field illumination can be routinely obtained by means of the critical and the Kohler methods (see Fig. 21), which are described in the following paragraphs.

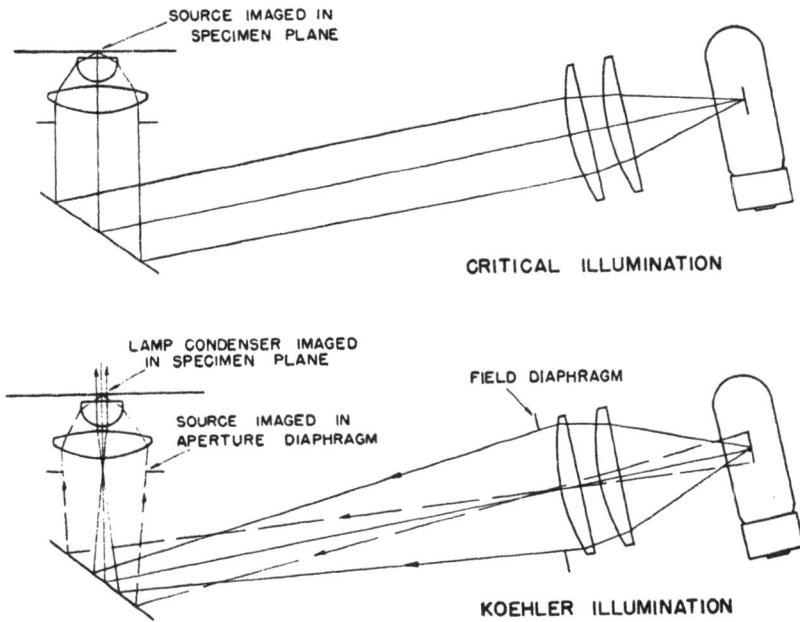

Fig. 21. Diagrammatic representation of critical illumination (top) *and Kohler illumination* (bottom).

APPROXIMATE CRITICAL ILLUMINATION

Critical illumination is basically achieved by focusing the image of the microscope lamp, that is, the ribbon of a filament bulb, via the mirror and the substage condenser on the specimen that is in focus. The lamp image, limited in size, is brought to a focus on only a narrow portion of the viewed specimen. This method gives rather uneven illumination.

An approximate approach to the critical method is most often used for routine microscopic procedures, since the finest results are not essential. Any source of diffused light from any standard microscope lamp can be used with this approximate approach, which is advantageous in that it can be set up rapidly with little adjustment and with no need for the light to be critically focused by the substage condenser. This approach gives adequate results in most cases.

WITH LOW-POWER DRY OBJECTIVES (10x)

It is assumed that the microscope is equipped with a substage condenser, an iris diaphragm, a mirror and a suitable source of light external to the microscope.

For approximate critical illumination, the following routine is suggested:

Step 1. The worker should become familiar with the microscope and its controls. The microscope can be used in a vertical position or tilted slightly so that the specimen can be viewed more comfortably. The microscope has coarse and fine adjustments for focusing. Turn these knobs and check whether they are working smoothly and properly. If a mechanical stage is part of the microscope, become familiar with its range of movement by moving it in both directions.

Step 2. Check the magnification power of the objective and the eyepiece. Choose the eyepiece (usually 10X) and the objective (10X) for moderate magnification. Raise the body tube—or lower the stage, if the microscope is equipped with a movable stage—and rotate the revolving nosepiece until the correct objective clicks into position.

Step 3. Place the specimen slide on the stage so that the portion to be viewed is approximately below the center of the objective. Secure the slide firmly by means of the stage clips. If a mechanical stage is part of the microscope, fit the slide gently into the mechanical stage and make sure that it is held firmly.

Step 4. Place the microscope illuminator in front of the microscope, with the light source on the same horizontal plane as the mirror and about 7 in. from the mirror. Switch on the illuminator current and turn the flat side of the microscope mirror toward the light.

Step 5. Adjust the substage condenser so that the top of the condenser is from 1 to 2 mm below the specimen slide. The substage condenser is provided with an iris diaphragm. Become familiar with it by rotating the lever (attached to the iris diaphragm) clockwise and counterclockwise several times. Adjust the iris diaphragm so that the aperture is opened approximately halfway.

Step 6. Tilt the mirror up and down or move it from side to side until the light is reflected upward through the optical axis of the microscope and a spot of light can be seen on the slide. Using the coarse adjustment, slowly lower the body tube—or raise the stage, if the latter is movable—until the lower part of the objective is about $\frac{1}{4}$ in. from the specimen slide.

Step 7. Place the eye as close as possible to the eyepiece and look down through the body tube. Keep both eyes open when looking into the microscope to reduce eyestrain. If eyeglasses must be worn while using the microscope, make certain that the glasses do not touch the eye lens or the metal mount of the eyepiece to prevent scratching the glasses and the eyepiece lens.

Look through the microscope and check whether the light is uniform throughout the field of view. If one part of the field appears brighter, adjust the mirror until uniform illumination of the whole field of view is achieved. Still looking through the microscope, use the coarse adjustment knob and either rack the tube up slowly or rack the stage downward until the specimen comes into view. Continue rotating the coarse-adjustment knob slowly in first one direction and then the other until the specimen is almost in sharp focus. Complete the sharp focusing of the specimen by slowly rotating the fine-adjustment knob.

Step 8. At this point, it is essential that the correct amount of illumination fill the objective. If the illumination is too intense, the glare may affect the resolution and sharpness of the specimen image. Remove the eyepiece and look through the body tube to observe the back lens of the objective, which should be uniformly illuminated; if it is not, tilt the mirror until it is. While still looking through the body tube, open or close the substage iris diaphragm until the illuminated image of the iris almost, but not quite, fills the entire back lens of the objective. Light filling about four fifths of the back lens area will usually be sufficient. Do not open the iris diaphragm so that the illumination spills over beyond the diameter of the back lens, since this creates more glare and less resolution.

Step 9. Still looking through the body tube, rack the condenser very slowly up or down several millimeters. If no change is observed in the illumination, the condenser is in the optimum position for approximate critical illumination (see Fig. 22). If the condenser has been racked down too far, the illumination on the back lens may be very uneven; therefore, rack the condenser up until the back is filled with uniform light.

Fig. 22. Diagrammatic representation of the correct iris diaphragm opening with the eyepiece removed.

Step 10. Finally, replace the eyepiece and again bring the specimen into sharp focus with the fine adjustment.

WITH HIGH-POWER DRY OBJECTIVES (40–43x)

The method of setting up the microscope for use with high-power dry objectives, where only a small portion of the specimen is to be examined, is similar to that for the 10X objective. It is assumed that the proper illumination has been arranged in the same manner as for the low-power objectives and that the area of the specimen to be examined has been chosen and centered on the stage. The following additional steps are then taken when a high-power dry objective is used:

Step 1. Rotate the nosepiece of the microscope carefully so that the objective clicks into position over the specimen. Caution is necessary at this point so that when the nosepiece is rotated, the objective is clear and does not touch the slide, since otherwise the objective lens and specimen slide may be damaged.

Most microscopes permit the 10X objective and the high-power dry objective to be safely interchanged by rotating the nosepiece. Also, most nosepieces carry objectives-designed to be parfocal and parcentric. This means that when the nosepiece is rotated to bring another objective in position, the center portion of the previously focused specimen will be brought automatically to an approximate focus with the new objective and will also be approximately centered. For microscopes that do not have these features 'and for workers who are fearful of damaging the slides or objectives, the safer method is to rack the body tube up or the stage down slightly (about 1 cm) before rotating the nosepiece. With the eye placed on the same level as the slide, bring the objective very slowly into position so that it is slightly clear of the specimen slide and does not touch it.

Step 2. Look through the eyepiece and, using the coarse adjustment, slowly rack the tube up or the stage down until the specimen is brought into rough focus. Now use the fine adjustment to bring the specimen into sharp focus. If the specimen does not appear in approximate focus on the first try with the coarse adjustment because of too-rapid movement of the coarse adjustment, bring the objective closer to the slide again without touching it and repeat the focusing procedure more carefully. (Many experienced workers use a reverse procedure by racking the body tube down or the stage up to save time. This procedure is dangerous, since even experienced microscopists sometimes jam the objective into the slides.)

Step 3. Remove the eyepiece and look through the body tube to view the back lens of the objective. It will be seen that the illumination, which uniformly covered the back lens of the low-power objective (10X), will now occupy a much smaller portion of the back lens. Open and close the iris diaphragm until the illumination fills about four fifths of the back lens. If the back lens cannot be filled uniformly with light even with the iris diaphragm full open, rack the condenser slowly upward until light fills the back lens. Now close the iris slightly until the back lens is almost uniformly filled with light.

If necessary, tilt the mirror slightly to obtain more uniform illumination.

Step 4.   Replace the eyepiece and bring the specimen into sharp focus with the fine adjustment.

Step 5.   If the light is too intense for comfortable viewing, do not close down the iris diaphragm, but insert a neutral-tint or opal-glass filter in the holder below the iris diaphragm.

WITH OIL-IMMERSION OBJECTIVES (90–100x)

Immersion Media

The immersion oils or fluids used with any oil-immersion objective should have refractive indices closely similar to the refractive index of the optical glass of the lenses and cover glasses, so as to eliminate optical distortion. The purpose of the immersion fluids (see page 18).

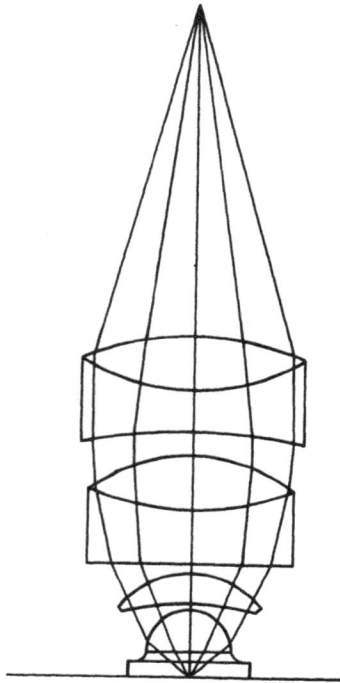

Fig. 23. *Diagrammatic representation of the path of light through an oil-immersion lens.*

is to allow the transmission of a larger cone of useful light through the objective (see Fig. 23). It will be remembered from the discussion of numerical aperture that the greater the cone of light entering the objective, the greater the resolution of the specimen and the observable detail.

Water was the earliest immersion fluid used, but since water evaporates too quickly, other types of immersion media are used today. Nondrying oils are generally used because they neither harden nor dry out during use. Oils that harden or dry may damage the objective lens. The most commonly used immersion oils are bought under proprietary brand names such as Shillaber or Crown oils. They are almost colorless, are nondrying, and have a refractive index of 1.5, about the same as that for optical glass. Refined cedarwood oil, formerly the most popular medium, can also be used, although it does thicken and harden with time.

Procedure

The procedure for setting up the microscope with the oil-immersion objective (97X) is similar in principle to that for dry objectives, except that an immersion fluid is placed on the top of the specimen slide, and the bottom portion of the objective is then immersed in this fluid. Great care must be observed in using the oil-immersion objective, since the working distance of this objective is very small.

Step 1. Set up the proper approximate critical illumination for the microscope as described for the high-power dry objective.

Step 2. Rack up the body tube or lower the stage, if the latter is movable, until the dry objective is substantially clear of the stage. Rotate the nosepiece until the oil-immersion objective clicks into position. Using a glass rod or small dropper, place a drop of immersion fluid on the slide below the objective. The exact point of application of the fluid is the circle of light formed by the microscope illuminator and condenser on the specimen slide.

Step 3. Lower the objective slowly with the coarse adjustment while simultaneously keeping the eye at the same level as the specimen slide until the objective lens is brought carefully into contact with the oil. This will cause the oil to spread evenly around the lens. Using the fine adjustment, continue lowering the objective very slowly and cautiously until the objective lens *almost* touches the slide.   Extreme care must be taken at this point not to jam the objective into the slide.

Step 4.   Open the substage iris diaphragm wide and look through the eyepiece for any indication of a specimen image.   If none is seen, move the slide slowly and smoothly, using care not to break the oil film, until any vague or blurred part of an image appears.   Using the *fine adjustment,* slowly raise the body tube or lower the movable stage until the specimen comes into view.   Focus slowly and carefully until the object is brought into sharp focus. If the oil film is broken during this procedure by moving the objective too far away from the slide, lower the objective again, carefully, into the oil and repeat the procedure as described.   Proficiency in bringing the specimen to a sharp focus with the oil-immersion lens will be gained by experience and good judgment.

Step 5.   Remove the eyepiece and examine the back lens of the objective. Open the substage iris diaphragm until the back lens of the objective is filled with uniform illumination.   It will be found that in most cases, the iris must be opened all the way.   If the back lens is still insufficiently illuminated, rack the substage condenser slightly up or down to increase the area of illumination.   Replace the eyepiece and observe the specimen.

For routine microscopic work, the procedure described above is adequate, but to reduce any stray light reflections between the top of the condenser and the bottom of the glass slide and to eliminate any loss of light due to refraction of the light passing from the condenser through air before reaching the slide, a drop of immersion fluid is also added to the top surface of the condenser.   Rack up the oiled condenser until the fluid is in contact with the bottom surface of the specimen slide.

Directly after use, clean the objective, condenser, and slides to avoid formation of hard films on the surfaces of the condenser and objective lens; such films are difficult to remove.   Moreover, the immersion oil, if left in contact with the objective, may seep through the sleeve of the objective and cause irreparable damage to the internal lens system of the objective.

KOHLER ILLUMINATION

The Kohler method, also considered a critical method of illumination, is most often used for achieving optimum results. It is described here in greater detail than critical illumination, since critical illumination is rarely employed and has become almost obsolete.

The Kohler method is most often used for high magnification or when uniform illumination of the specimen and the entire field of view is desired, so as to achieve the highest possible resolution and the best image quality. It is the recommended system for uniform illumination of a microscopic specimen, especially in photomicrography. Since this system requires additional adjustments if exact illumination is to be achieved, it is seldom used by novices. Once mastered, it is highly rewarding to the microscopist, since finer detail and a sharper, clearer image result.

Basically, with Kohler illumination the image of the light source or, to be precise, the coiled filament of the microscope lamp, is focused by a condenser attached to the microscope lamp onto the substage condenser of the microscope. (In practice, the light is actually focused on the substage iris diaphragm.) The substage condenser, in turn, brings the image of the lamp condenser into focus in the plane of the specimen, and the lamp condenser actually becomes the source of illumination.

The practical advantage of this system is that the specimen is uniformly illuminated without glare, even though the nonuniform brightness of a coiled filament source is used.

PROCEDURE

The microscope illuminator selected must be the advanced type equipped with a lamp bulb containing a ribbon filament, an adjustable focusing condenser (field condenser) to focus the image of the bulb filament, an iris diaphragm (field diaphragm) to control the size of the illuminated area, and a filter holder.

The lamp bulb in most illuminators is already precentered. If it is not, the lamp filament should be centered with respect to the optical axis of the field condenser of the illuminator by projection.

The more advanced and modern research microscopes have a source of illumination built into the base. These are designed to give a modified type of Kohler illumination and require only slight adjustment of the substage condenser and iris diaphragm. Intensity can be controlled by a transformer or glass filters.

The Kohler method consists basically of focusing the image of the lamp filament on the substage condenser of the microscope. The substage condenser then focuses this image in the plane of the specimen after which the substage iris diaphragm is opened sufficiently to almost fill the back lens of the objective with uniform light. In actual practice, the lamp image is brought to focus on the substage diaphragm of the microscope and not on the substage condenser (see Fig. 24).

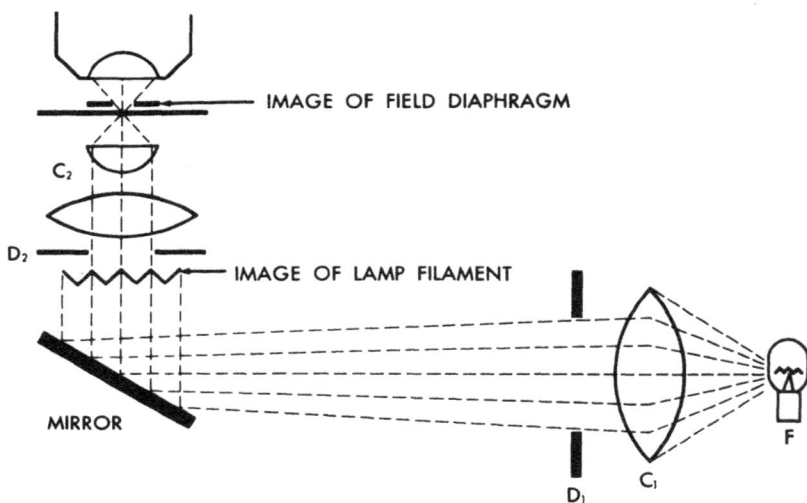

Fig. 24. *Diagrammatic representation of Kohler illumination (described in the text). Steps 1–4 : An image of the lamp filament is formed in the plane of the substage diaphragm* (D$_2$) *by focusing the lamp condenser* (C$_1$). *Steps 5 and 6 : An image of the field diaphragm* (D$_1$) *is formed in the specimen plane by focusing the substage condenser* (C$_2$). *Step 7 : The opening of the substage diaphragm* (D$_2$) *is adjusted to fill the back lens of the objective when viewed in the microscope tube.*

The steps for achieving Kohler illumination follow :

Step 1. Position the microscope illuminator so that the iris diaphragm of the illuminator (field diaphragm) is from 6 to 9 in. from the substage mirror.

Step 2. Tilt and adjust the mirror so that the light from the lamp is centered in the plane side of the mirror and reflected upward through the optical axis of the microscope.

Step 3. Close down the substage iris diaphragm of the microscope. Move the variable focusing condenser of the illuminator backward and forward to vary the distance between the lamp filament and this field condenser until a sharp lamp image is projected on the back surface of the substage iris diaphragm. If the condenser is not movable, move the lamp itself to obtain the focus. Focus the image of the lamp filament as sharply as possible on the iris. The image should be slightly larger than the diameter of the iris when it is fully opened. If the housing of the illuminator is supplied with a ground glass plate, remove the plate temporarily during the various adjustments of the lamp and then replace it. If the bulb supplied with the illuminator is frosted, attempt to focus the frosted part of the bulb on the iris diaphragm.

Step 4. Adjust the mirror while focusing, so that the filament-bulb image is centered in the mirror. The focusing of the filament image on the iris may be aided by viewing the image directly in the microscope mirror. Another method is to view the substage mirror image in a mirror held in the hand.

Step 5. Open the substage iris diaphragm to allow adequate illumination of the specimen. Look through the eyepiece and bring the specimen into focus. Now close down the lamp iris diaphragm (field diaphragm) to a small aperture. Look again into the microscope and rack the substage condenser up or down until a sharp image of the iris-diaphragm aperture is focused on the specimen plane. If necessary, tilt the mirror and adjust the position of the illuminator slightly, so that the image of the field-diaphragm opening is centered in the mirror.

Step 6. Looking again through the microscope, open the field iris diaphragm slowly until the image of its opening is equal or slightly greater than the microscope field of view. This adjustment will illuminate the specimen sufficiently for good observation. Too much light may cause glare and impair the resolution of the viewed image.

Step 7. Remove the eyepiece and examine the back lens of the objective. Adjust the substage iris diaphragm so that the transmitted light will almost fill the objective. Generally speaking, if about 3/4 or 4/5 of the diameter of the back objective lens is illuminated, maximum resolution of the specimen will be achieved.

Step 8. If the light passing through the microscope and reaching the eye is too intense for comfortable viewing, *do not* cut down the intensity of the light at this stage by closing down the substage iris diaphragm. Rather, decrease the light by placing a neutral density filter in front of the illuminator. The illuminator usually has a filter holder for this purpose. The explanation for this step is that any additional manipulation of the microscope adjustments or illuminator adjustments will interfere with proper achievement of Kohler illumination. Any manipulation of the substage iris diaphragm to control the intensity of the light will cause the back lens of the objective not to be properly filled with light and the optimum resolution of the specimen will not be achieved.

Oblique Illumination

This type of illumination is useful for the microscopic study of special specimens, such as various species of diatoms. At times, it is used with reflected light to detect uneven surfaces of objects, but great care must be taken when evaluating viewed specimens under oblique illumination, since oblique light may cause color fringes or radical changes in the pattern of the viewed object.

PROCEDURE

To obtain oblique illumination, swing the mirror to one side so that it is off-center. The light will then pass through the condenser from only one side and will illuminate the specimen by a cone of light that is at an oblique angle. (Many advanced microscopes are provided with a substage attachment and control knob for oblique illumination, making it possible to obtain any angle of-obliqueness of light through an arc of 90° by a corresponding movement of the control knob attachment.)

Dark-Field Illumination

When viewing a specimen under the microscope, the human eye is limited by what it perceives in the differences or contrasts of brightness and color within the specimen itself or between the specimen proper and its background. For the human eye actually to see anything, these two characteristics—contrast in brightness and in color—must be present in the object. The color of the object, as seen by the eye, depends on the wavelength of the light emanating from the object; the brightness depends on the amplitude of these light waves. The function of the microscope, then, is to magnify an object in such a manner as to cause difference in brightness and color, so that the eye will be able to see details that reveal the nature of the object.

Most specimens, as viewed normally with bright-field illumination, appear dark against a bright background and give sufficient contrast so that the details in the specimen are adequately resolved. In those cases where the object under view lacks contrast in color and brightness, however, many procedures and devices, such as phase-contrast, interference, and polarizing are employed to enhance the effectiveness of the microscope. One such procedure is dark-field illumination.

The dark-field method, which gives concentrated oblique illumination, permits minute particles of the specimen, too small to be ordinarily visible, to scatter the light and thus become visible to the eye. This effect is similar to that experienced when dust in a darkened room is seen in a beam of sunlight or when smoke particles are observed floating in a beam of strong light. The more intense the illumination against the darkened background, the smaller the particles that can be perceived, since the eye is most sensitive under these conditions. With dark-field illumination, the background appears dark and the specimen appears bright, thus offering intense contrasting effects. This method of illumination is used extensively in medicine as a diagnostic tool for evaluating small organisms such as bacteria and microorganisms.

To obtain dark-field illumination, a special field stop (see Fig. 25) is placed below the condenser, in the filter holder, or a special dark-field condenser (paraboloid or cardioid) is used. A simpler,

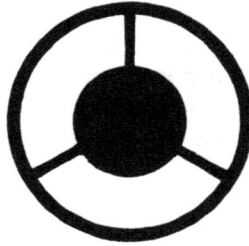

Fig. 25. Dark-field diaphragm.

though less efficient, device is a circular piece of black paper, about half the diameter of the top lens of the condenser, placed at the center of the top lens of the condenser. When the substage iris diaphragm is opened wide, the eyepiece is removed, and you look down the body tube, no direct light should enter the objective in the absence of a specimen.

The dark-field stop blocks the direct light from the midsection of the condenser and thus keeps it from reaching the objective. The light, however, is allowed to pass through both ends or margins of the condenser to illuminate the specimen. The oblique or marginal rays of light are such that although they illuminate the object, very little light enters the objective but rather passes from the field of view. The object thus appears self-luminous against a dark background.

The following few rules must be known before one can attempt dark-field illumination:

1. A very intense, nondiffused light source is essential for good dark-field illumination. The ordinary substage microscope lamp will not give sufficient illumination. A 100- or 200-watt coil-filament bulb or the light from a carbon-arc lamp will be adequate.

2. All adjustments necessary for dark-field illumination are first accomplished with the 10X low-power objective; then the change is made to the 40X or the oil-immersion objective.

3. The specimen to be examined must be placed on a slide either 1.15 or 1.25 mm thick. (Special dark-field condensers require and generally specify slides of a specific thickness, since otherwise the light will not be properly focused on the specimen.)

4. Slides must be thoroughly clean, free from all dirt or other foreign matter. Any potential increase in contrast will be destroyed if dirty or smudgy slides are used.

5. The specimen or preparation to be examined under dark-field illumination must be thin and not dense in order to avoid extraneous scattering of the light.

6. The specimen should be mounted in a medium with a higher index of refraction than air.

7. There must be no bubbles in the mounted specimen or in the immersion oil.

8. When an oil-immersion objective is used, a drop of oil should be placed between the top of the dark-field condenser and the glass slide to achieve optimum efficiency.

9. The special dark-field condenser, if used to replace the normal condenser, must be accurately centered, especially when specimens are to be viewed with oil-immersion objectives. These special condensers are generally supplied with centering screws.

SPECIAL DARK-FIELD CONDENSERS

The special dark-field condensers (see Fig. 26) that are commercially available generally have a numerical aperture of 1.2 to 1.4. The N.A. of the objective used in conjunction with the dark-field condenser must be lower than that of the condenser; otherwise, the objective may accept some of the transmitted light and impair the contrast. The N.A. of the objective is usually reduced by means of a barrier called a *funnel stop,* which fits into the back of the objective, although this method is not very efficient. Some objectives have built-in iris diaphragms for adjusting the N.A. The N.A. of the objective used in conjunction with a dark-field condenser having an N.A. of 1.20 to 1.40 must be reduced to about 1.0. For condensers with apertures of about 1.00, the N.A. of the objective is reduced to about 0.85.

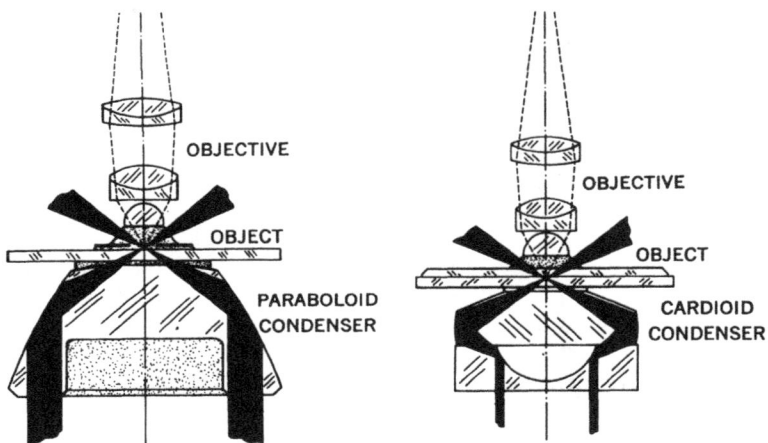

Fig. 26. Diagrammatic representation of dark-field condensers : (left) *paraboloid ;* (right) *cardioid.*

PROCEDURE

The general procedure for dark-field illumination using an oil-immersion objective is described below. (The procedure when using dry objectives with dark-field condensers is similar.)

Step 1. Place a blank glass slide on the stage of the microscope. Remove the bright-field condenser and insert the dark-field condenser in the correct position, with its two lateral screws in the forward position. Rack up the condenser so that it just touches the bottom of the glass slide.

Step 2. Position the illuminator or arc lamp about 8 in. from the substage mirror, with the plane side facing the source of illumination. If filters must be used, use only transparent filters to avoid any diffusion of the light. Center the beam of light in the mirror so that it is reflected upward toward the dark-field condenser. Rotate the nosepiece until the 10X objective is in position. The lower-power objective is used primarily to bring the desired portion of the specimen into focus and to center the condenser.

Step 3. Carefully rack either the body tube downward or the stage upward, and bring the centering ring of the dark-field condenser into focus. The ring, which is generally etched on the surface lens of most dark-field condensers, appears as a luminous ring against a dark background. If the bright ring is not in the middle of the field, center the condenser by turning the centering screws or by any alternate method described by the manufacturer. Adjust the mirror carefully until the luminous ring appears at its brightest.

Step 4.   Remove the blank glass slide from the stage. Place a large drop of immersion oil on the top lens of the condenser and lower the condenser slightly. Place the specimen slide and cover glass on the stage and raise the condenser slowly until a good oil contact is made between the condenser and the bottom of the slide. No air bubbles must be entrapped in the oil or the specimen image will be impaired.

Step 5.   Using the 10X objective, bring the desired portion of the specimen to a focus. Manipulate the slide so that this portion is brought to the center of the field of view.

Step 6.   Raise the body tube, or lower the stage slightly. Swing the oil-immersion objective into its working position. Reduce the numerical aperture of the objective sufficiently by using either a funnel stop or the iris diaphragm that is part of the special oil-immersion objective used with dark-field illumination. If a funnel stop is used, unscrew the lower half of the oil-immersion objective and insert the funnel stop with the small end toward the lens.

Step 7.   Place a drop of oil, free from bubbles, on the cover glass. Using the fine adjustment, bring the oil-immersion objective slowly and carefully into contact with the cover glass, so that the oil is spread evenly. Focus very carefully with the fine adjustment until the specimen is in perfect focus.

The organisms should appear as bright objects against a dark background. Adjust the source of light so that the specimen is brilliantly illuminated. If the light is too intense for viewing, reduce it by adjusting the condenser diaphragm.

Ultraviolet Illumination

Ultraviolet light is used in photomicrography to improve the recorded image of the specimen, since a reduction in the wavelength of the light will lead to a greater increase in the resolution of fine detail. The resolution is increased about twofold when ultraviolet illumination is employed, permitting observation of details that would be invisible under normal examination.

Since ultraviolet light is invisible to the human eye, it is used mainly in still or motion-picture photomicrography. (*Ultraviolet radiation is very injurious to the eyes and skin, so precautions must be taken at all times to shield the observer from stray radiation.*)

The mercury-vapor arc lamp, with the correct filter, can be used as a source of ultraviolet light (see Table 5). The transmitted light will be in the upper region of the ultraviolet spectrum (365 nanometers). Since optical glass transmits a large proportion of ultraviolet light, standard apochromatic objectives, slightly corrected for chromatic aberration in this region, can be used with a compensating eyepiece.

PROCEDURE

The procedure for using ultraviolet light (365 nanometers) with the standard microscope is as follows.

Step 1. Switch on the mercury-vapor arc lamp, which emits both visible and ultraviolet radiation. Place a Wratten 77A filter in the filter holder. This filter transmits 68% of the visible green light from the mercury lamp and thus permits the specimen to be visibly focused.

Step 2. Place the specimen on the stage and rack up the condenser until it almost touches the slide. Open the substage iris diaphragm wide and bring the specimen to an approximate focus.

Step 3. Close down the substage iris diaphragm and follow the procedure described for the Kohler system of illumination.

Step 4. Bring the specimen to a fine focus, using the fine adjustment.

Step 5. Replace the Wratten 77A filter with a Wratten 18A filter. This filter is extremely deep red in color which absorbs the visible light and transmits the ultraviolet radiation particularly in the 365-nm wavelength region.

Step 6. Record the image of the specimen on film (see Chapter 11).

USING THE LOWER ULTRAVIOLET RANGE

If the full range of ultraviolet light, derived from a special cadmium spark lamp, is to be used (wavelength 275 nanometers), quartz optics must be used, since ordinary optical glass will absorb most of the ultraviolet radiation in this spectral range.

Immersion-fluid quartz objectives can also be used with ultraviolet light. A solution of cane sugar and glycerin, which is non-hygroscopic, can be used as the immersion fluid. This mixture will transmit the ultraviolet radiation and have a refractive index closely approximating the quartz objective (1.496) when using radiation in the wavelength range of 275 nanometers or lower.

To avoid eye injury from the dangerous invisible rays, preliminary visual focusing is not done. The focusing is performed by placing a photomicrographic camera provided with a ground-glass focusing screen over the eyepiece of the microscope and then substituting a special uranium-glass plate for the ground-glass screen. The image will fluoresce on the uranium glass plate and can then be brought into focus.

Another method of bringing the specimen into focus is by trial and error. The focus of the specimen in ultraviolet light is determined by a series of test exposures with the photomicrographic camera at different focal levels, which are read from the graduated fine-adjustment knob. Choose the focal level that gives the sharpest picture.

FLUORESCENT MICROSCOPY

When examined under ultraviolet radiation, many materials *fluoresce*; that is, excited by the ultraviolet radiation, they emit light and become visibly colored against a dark background. The fluorescent color visible to the eye is usually yellow to green (to which the eye is very sensitive), although red or blue fluorescence is actually present but masked by the stronger greenish color.

Fluorescence is primarily a function of chemical composition, and certain conclusions can therefore be drawn as to the chemical composition of a specimen under fluorescent light. Fluorescent microscopy is used to delineate certain structures that are not visible with other types of illumination. Specimens can be treated with substances that fluoresce. These fluorescent materials have a selective affinity for different parts of the specimens, whether stained or unstained, and this preferential identification of materials that show selective fluorescence can therefore be used to determine their chemical composition and characteristics. Such selective fluorescence is widely used in pathology and medicine.

Certain substances called *fluorochromes,* which are not necessarily stains, have a specific affinity for cellular structures and tissues. Among other uses, they serve to differentiate types of large viruses and in leucocyte counts. Fluorochromes can be classified as acidic, basic, or neutral. Examples of each are: basic—auramine, acridine orange, revanol, berberine sulfate; acidic—thiazine red, fluorescein, sulforhodamine, primuline; and neutral—rhodamine B.

The technique employed in fluorescent microscopy is the same as that for ultraviolet illumination. Two kinds of filters are used—those that transmit only the ultraviolet radiation· from the light source to the specimen, and barrier filters which filter out the ultraviolet radiation coming from the specimen and allow only visible light to reach the eye.

PROCEDURE

Using the mercury-vapor arc lamp and a Wratten 77A filter, bring the specimen into sharp focus. Replace the 77A filter with a Wratten 18A filter and insert a Wratten 2B filter disc in the eyepiece. The 2B filter absorbs the ultraviolet radiation and transmits only the visible portion of the spectrum.

In fluorescent microscopy, it is essential that the specimen be illuminated only with ultraviolet light. Beyond the specimen, only visible light is necessary for viewing the specimen and for photomicrography.

The standard objectives and eyepieces can be used in fluorescent microscopy.

Infrared Illumination

Infrared illumination is sometimes utilized to accentuate the contrast of details within the specimen. Special stains absorb the infrared radiation and thus increase contrast. Infrared light, which is composed of longer wavelengths, is also used to make dense specimens, opaque to normal illumination, translucent. It must be noted that because of its longer wavelength, infrared light will cause a loss in resolution. Infrared radiation is achieved by using normal illumination in conjunction with selective filters. Since the rays are invisible, they are useful chiefly in photomicrography.

PROCEDURE

Bring the specimen into focus under Kohler illumination and insert an infrared filter in the filter holder. Since infrared radiation is invisible to the human eye it must be recorded on special film sensitive to infrared light.

Focusing for photomicrography is achieved by trial and error, as with ultraviolet radiation; the correct focus is determined by a series of test exposures with a photomicrographic camera at different focal levels read from the graduated fine-adjustment knob. Choose the focal length that gives the best recorded image.

LOCATING OBJECTS IN THE FIELD

A straight hair can be used as a pointer in the following manner: Place the hair across the middle portion of the diaphragm located within the eyepiece. Cement both ends of the hair down and allow the cement to harden. Cut the hair just to one side of center and remove the longer piece with its cement. The attached hair acts as a pointer and locates any portion of the slide when rotating the eyepiece and moving the specimen slide.

Certain points on a slide can be marked to facilitate locating them for future reference. One method of doing this is as follows: After focusing on the specimen with the low-power objective (10X), raise the objective and, with ink, draw a small circle on the cover glass around the specimen situated over the opening in the stage. Bring the specimen into focus again and determine the position of the desired part of the specimen within the circle. Raise the objective once more and draw another, smaller circle around the part of interest. When the ink is dry, raise the objective and remove the slide; on the reverse side of the slide, trace the smaller circle in ink and, when the ink is dry, remove the original circles on the cover glass with a moistened cloth. The lower circle isolates the desired part of the specimen.

Another method is to use a mechanical stage on which two scales are engraved at right angles to each other; these scales give two coordinates for locating any desired point. The specimen is brought to the center of the stage and with the image in focus, the position of the desired part is read on the two measuring scales and a record made of the slide and the coordinates.

Still another method is to use a *finder slide* (see Fig. 27). Finder slides, several types of which are commercially available, are ruled

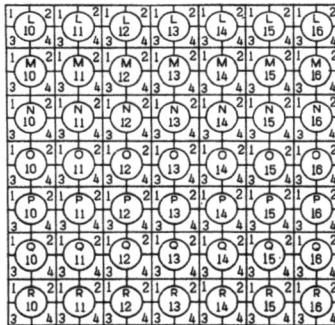

Fig. 27.   Finder slide.

with rows and columns of numbers or letters that have been reduced photographically. The specimen slide is centered on a mechanical stage and brought into focus with the desired portion in the center of the field of view. The slide is removed and the finder slide is mounted in the same position in the mechanical stage, care being taken that the mechanical stage is not moved while replacing the specimen slide. The position on the finder slide that is centered in the field of view is then noted; with the type of finder slide illustrated, it might be, for example, *P15-3* (the last digit denoting the quadrant). To find this part of the field again, the finder slide is mounted in the mechanical stage and position *P15* of the slide is centered in the field of view.   The finder slide is then removed and the specimen slide placed in this exact position.

If a mechanical stage is not available, vertical and horizontal baselines for aligning slides can be made temporarily on the stage with a crayon or pencil; they can also be permanently scratched with a small needle and filled in with a colored crayon.

MAKING DRAWINGS OF MICROSCOPIC OBJECTS

The best method of obtaining an accurate reproduction of a specimen is through the use of a photomicrographic camera (see Chapter 11 for a description of this process).   This method, however, requires practice and additional instrumentation.   An easier method of obtaining a reproduction of a specimen is to project it on a sheet of paper and trace it.   Microprojectors are manufactured in various styles and sizes.

Using the Camera Lucida

The camera lucida (see Fig. 28) is a special device for aiding the microscopist to make drawings of microscopic specimens; it is

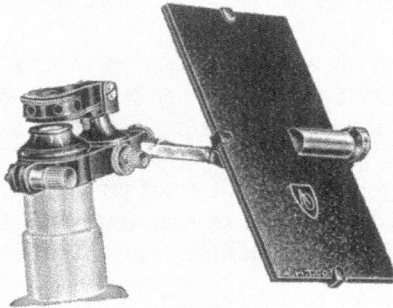

Fig. 28.   Camera lucida.

attached to the top of the body tube of the microscope, directly over the eyepiece.   It consists basically of a prism positioned directly over the eyepiece.   A variety of neutral-tint filters can be set in position by means of a rotatable mount. These filters serve to equalize the light coming from the eyepiece and the light coming from the drawing paper; such balancing of the light is necessary for good drawings. A large mirror, attached by a side bar to the side of the body tube, can be inclined to any desired angle.

To make a drawing, the paper is placed under the mirror and the light is balanced by the neutral filters. By looking through the prism, the observer can see both the drawing paper and the specimen. By holding his pencil over the paper, he can see the movement of the pencil as if it were under the objective lens of the microscope and thus make drawings with ease and accuracy.

Making A Rough Drawing

When no device for aiding the microscopist is available, a rough drawing of a specimen can be made in the following way: Place the drawing paper on a raised platform about 10 in. below the eyepiece and at the right side of the microscope. The plane of the paper will then be parallel to the plane of the magnified virtual image. With both eyes open, it is possible to trace the outlines of the specimen with a pencil, with no conscious effort to follow its movement.

CARE OF THE MICROSCOPE

The microscope is a delicate instrument and must be handled with the utmost care at all times. When it is not in use, keep it covered with a plastic cover or a glass bell jar, or keep it in its case.

Store it in a place where it is not subjected to extremes of temperature, which may expand and contract the glass of the lens with changes in temperature and thus cause separation of some of the lens elements.

The microscope must be kept free from dust and dirt, its major enemies. Any dust that collects on such mechanical parts as the rack and pinon gears can be wiped off with a soft linen cloth, followed by brushing with a camel-hair brush.

Never use oil to lubricate the movable parts, since only a special type of oil is sparingly used. Lubricating oils used indiscriminately may deposit gums and greases within the mechanism and cause malfunction of or damage to the movable parts.

The mirror, the upper and lower surfaces of the eyepiece, and the bottom surfaces of the dry and oil-immersion objectives should be cleaned regularly. Extreme care must be taken when cleaning the optical parts of the microscope, however, since the lenses may be easily scratched in the process. Do not use any cleaning agents containing abrasives or grit.

If there is dust on the optical parts, remove it by brushing very softly with a camel-hair brush or by blowing dry air over the lens with a rubber bulb. If the optical parts are dirty or greasy, clean them with soft, lintless lens tissue. If the dirt is resistant, wipe gently with a soft linen cloth, using the least amount of pressure to do the job. Do not use chamois skins, since they may contain hard particles, grit, or residual oils. The routine method of cleaning the glass surfaces is to breathe on the optical part to deposit a small amount of moisture and then wipe clean before the moisture evaporates. Repeat this procedure several times until the optical part is clean.

If an accumulation of grease does not respond to this treatment, moisten the soft linen cloth with alcohol and wipe gently over the lens surface. Repeat until the grease is removed. Dry immediately with a piece of lens tissue and remove any lint with a camel-hair brush.

If the grease does not respond to the alcohol treatment, xylol, on a moistened linen cloth, can be used sparingly as a last resort. Care must be observed in using this solvent, since xylol can penetrate the sleeve of the objective and dissolve and weaken the cementing material that holds the various lenses within the objective in place.

Never allow immersion oils to remain for long periods on the objective. Wipe clean immediately after use, then use a little xylol to remove any remaining oil, and finally remove the xylol with lens tissue.

Never attempt to take the objective apart for cleaning. If the optical parts need repairing, return them to the factory.

CARE OF THE EYES WHEN USING THE MICROSCOPE

It is essential that the eyes be used most judiciously when using the microscope so as to avoid any strain or impairment of the eyesight. The following simple rules, if scrupulously observed, will allow the comfortable use of the microscope over long periods of time :

1. The position of the eyes for most efficient viewing is just above the eye lens.

2. If the microscope is of the monocular type, keep both eyes open when looking through the microscope, alternating each eye at the eyepiece. Do not squint or close the free eye. This procedure may require a little practice at first, but use of both eyes will cause less fatigue and permit longer use of the microscope.

3. Do not peer through the microscope for long periods or until the eyes become excessively fatigued. Rest the eyes from time to time.

4. If you wear glasses, it is better to remove them when looking through the microscope. The coarse and fine adjustments compensate for such eye defects as nearsightedness or farsightedenss, and if you keep your glasses on, they may be scratched, chipped, or broken by the metal rim of the eyepiece.

If glasses must be worn because of an astigmatic condition, apply a rubber guard or soft material around the mount of the eyepiece. Such a guard can be purchased, or one can be made by cutting off the top of a rubber finger and pulling the open end over the eyepiece until the eye lens of the eyepiece is just exposed. A thick rubber band can also be used for the same purpose.

5. Avoid direct sunlight or light that is too bright or too dim.

CHAPTER 3

The Stereomicroscope

THE STEREOMICROSCOPE has become a very important tool in industry for assembling, examining, and inspecting small parts. It has become a very useful instrument for studying biological, pathological, and mineralogical specimens. It has also been used extensively for dissections of biological specimens, where the specimens are seen undistorted and in depth.

The major benefit derived from stereoscopic magnification is the three-dimensional effect. Objects viewed with a stereomicroscope are magnified in three dimensions, with the specimen upright and unreversed and therefore appearing realistic. Stereomicroscopy does suffer from a major disadvantage, however, in that the total magnification that can be achieved with present-day stereomicroscopes is about 150X. Observation of very fine detail is therefore limited by the use of stereomicroscopy.

The stereomicroscope is basically two compound microscopes, since its optical system includes two sets of objectives and two eyepieces (see Fig. 29). With this setup, the stereomicroscope forms two slightly different images of the specimen. Each eye, looking through one of eyepieces, sees one of these images. The brain combines these two images, thus giving to the viewer the illusion that the specimen is seen in depth.

The high-power stereomicroscopes currently available, however, use only one objective. A prism placed in the focal plane of the objective gives two views of the same image of the specimen. Half of the objective is used to diffract the light and form an image for the left eye, while the light diffracted from the other half forms a slightly altered image for the right eye.

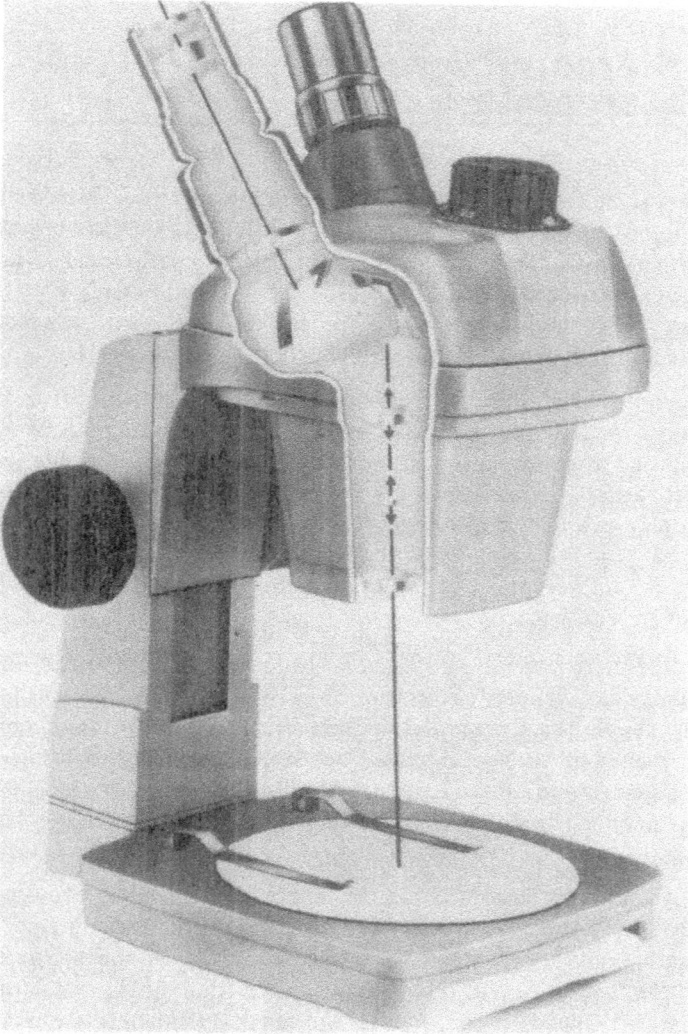

Fig. 29. Stereomicroscope.

The standard stereomicroscope actually consists of two microscopes mounted together. Each microscope consists of an objective, an inclined eyepiece, and a prism positioned behind the objective. The prism is mounted on a synchronized bearing to provide a means of adjusting the interocular (that is, the distance between the eyes) necessary for the observer without affecting the focus of the specimen. The angle between the two microscopic axes for viewing the specimen is usually about 10°, which is about the same angle as the convergence of the eyes. This angle is sufficient to obtain a stereoscopic view of a specimen at a distance of about 14 in.

The paired eyepieces used in the stereomicroscope are the "wide-field" type, which give at least a 25% greater field of view. The eyepieces are generally corrected for distortion and are made in 10X, 15X and 20X power.

The paired stereoscopic objectives are achromatic. When they are enclosed permanently with the prism system into a single unit, they are called *power pods*. When using different magnifications, it is possible to mount different power pods into the stereomicroscope or to change the eyepieces having different magnifications.

Stereomicroscopes are also commercially available with a "stereo-zoom" feature. Different magnifications within a limited range, dependent on the power of the objectives and eyepieces, can be achieved by simply turning a dial. The image will remain in focus through any stage of magnification.

No condenser is used because of the low magnifications and the small numerical apertures of the low-power stereoscopic objectives. The condenser generally adds very little to the resolution of the specimen when used in conjunction with objectives with low numerical apertures.

The stage is a clear glass plate, free of bubbles and striations, for transparent specimens. The glass plate can be removed easily and replaced with an opaque opal glass plate, black on one side and white on the other, for opaque specimens, or the plate can be made of rigid metal.

The stand that supports the various optical parts is rigid and generally constructed of metal. The sole control for focusing adjustment, on the arm of the stand, is actuated by a rack and pinion.

The base of the stand below the stage is equipped with a concave mirror, which can be rotated. It can be easily replaced with an illuminator inserted into the base of the stand. An illuminator can also be inserted in the arm of the stand so that incident light can be obtained for viewing opaque objects.

USING THE STEREOMICROSCOPE

With a few differences the use of the stereomicroscope is similar to that of the standard compound microscope.

1. Transparent specimens are placed on the clear glass plate, which acts as the stage. The working distance for most stereomicroscopes is 4 in., except when special attachments are used. The long working distance is useful for observing large specimens.

2. The light from the microscope lamp is reflected by the microscope mirror to illuminate the specimen. Since low magnifications are used, the method of illuminating the specimen is not critical. Diffused light to illuminate the specimen adequately is sufficient. If the source of light is too intense, dim the illumination by means of neutral-density filters or by increasing the distance between lamp and mirror. Some stereomicroscopes are equipped with a built-in illuminator that eliminates the mirror and transmits sufficient light directly to the specimen.

3. When the specimen is opaque, the clear glass plate is removed from the stage and replaced with the matte-ground opal glass plate. Either the black or white side is selected, depending on which gives the better contrast.

A beam of light from above is directed onto the specimen at an angle to illuminate the specimen adequately. Generally, a more intense source of light is necessary to illuminate opaque specimens. Therefore, special stereoscopic illuminators are available which can be adjusted by means of a transformer to give brilliant, intense, cool light over small areas of the specimen. The sharpness and clarity of the stereoscopic image will depend largely on the type of lighting and its intensity.

4. It is essential that the stereomicroscope be aligned properly so that the two images are blended into one and seen by the eyes as a crisp three-dimensional image. The interpupillary distance must therefore be adjusted by the observer so that only one image is seen; otherwise, eyestrain will occur. If, after adjusting the interpupillary distance, a double image is still seen, the microscope is defective and should be returned to the vendor for repair.

CHAPTER 4

Measuring and Counting with the Microscope

MEASURING WITH THE MICROSCOPE

Two accessories are needed for measuring objects under the microscope. The first is the *stage micrometer slide*, which is a glass slide, 25 by 75 mm, with an accurate fine scale ruled or mounted on it (see Fig. 30). The scale is 2.2 mm long, the first 0.2 mm being divided into 0.01 mm and the remaining 2.0 mm into 0.1 mm graduations. The scale on many stage micrometers is protected by a cemented cover glass.

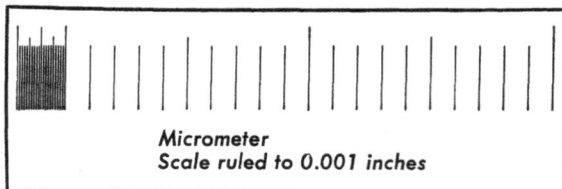

Micrometer
Scale ruled to 0.001 inches

Fig. 30.   Stage micrometer slide.

The second accessory is the *eyepiece micrometer disc reticule*. There are many varieties of micrometer discs available with different ruled scales designed for special purposes. The standard type, however, has a micrometer disc 21 mm in diameter, with a 5-mm scale ruled on the disc, the scale being divided into 50 parts of 0.1 mm each (see Fig. 31).

The top lens of the eyepiece is unscrewed and the micrometer disc

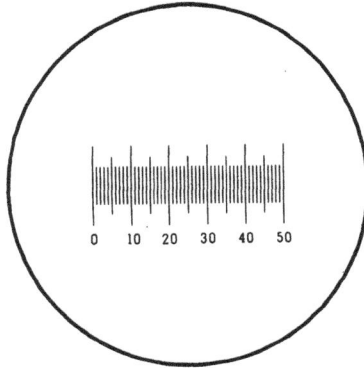

Fig. 31.   Eyepiece micrometer disc reticule (5-mm scale).

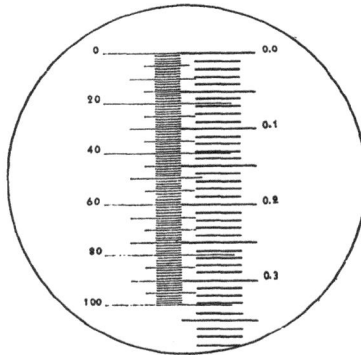

Fig. 32.   Scale of stage micrometer slide superimposed on scale of eyepiece microme-ter disc.

is placed with the ruled side down on the diaphragm within the eye-piece.   The eye lens is then screwed back.

The apparent length of each scale division of micrometer disc depends on the total magnification to be used.   Consequently, the scale of the disc must be calibrated for each combination of eyepiece and objective used with the microscope.   If the microscope in use has an adjustable body tube, the calibration of the eyepiece micrometer disc must also include each setting of the body tube.

Calibration Constant

The procedure for calibrating the scale of the eyepiece micrometer disc reticule is as follows :

Step 1.   Place the stage micrometer slide on the microscope stage. Look through the eyepiece and bring the scale of the stage micrometer slide into sharp focus.   The image of this scale will appear superimposed on the scale of the eyepiece micrometer disc.

Step 2.   Move the stage micrometer slide until its zero line corresponds exactly with the zero line on the scale of the eyepiece micrometer disc (see Fig. 32). The dimension of each division of the eyepiece micrometer scale can be determined in terms of the true scale of the stage micrometer for this specific magnification.   It is necessary only to determine how many divisions or parts of a division of the eyepiece micrometer scale equal one division on the scale of the stage micrometer.   The ratio of the numerical value of the stage micrometer scale and the numerical value of the eyepiece micrometer scale is called the *calibration constant* of the eyepiece micrometer.   Once determined for a specific combination of eyepiece and objective, this calibration constant can be used subsequently without repeating the calibration procedure.

For example, if A is the true length seen on the stage micrometer scale and corresponds to B, the number of divisions seen on the eyepiece micrometer scale, then the calibration constant will be A/B. The size of an object can now be measured with the eyepiece micrometer alone by multiplying the size of the object, as measured on the eyepiece micrometer scale, by the calibration constant.

After calibrating the eyepiece micrometer, remove the stage micrometer and place the specimen to be measured on the microscope stage. Determine the number of divisions of the eyepiece micrometer that measure the specimen or part of the specimen.   Determine the true size by multiplying by the calibration constant.

Filar Micrometer Eyepiece

The Filar micrometer eyepiece (see Fig. 33, *top*), a special type of Ramsden eyepiece, can be used to measure surface defects, surface

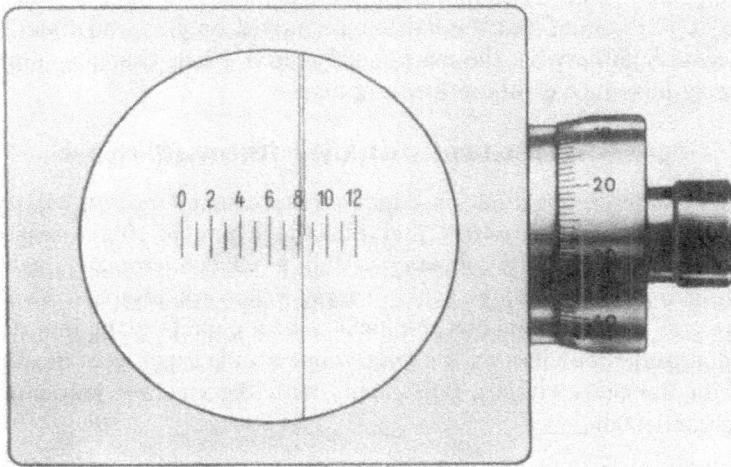

Fig. 33. Filar micrometer eyepiece (top) *and filar micrometer eyepiece scale* (bottom).

irregularities, and size of specimens. The eyepiece is equipped with two hairlines and a micrometer scale, which are positioned within the eyepiece in such a manner as to be superimposed on the surface of the focused image.

The eye lens of the Filar eyepiece can be adjusted so that the hairlines and micrometer scale can be accurately focused on the image of the specimen. One hairline is fixed across the center of the field to act as a guide or reference line to the second cross hair, which is moved by means of a graduated micrometer knob or drum at the side of the eyepiece. The circumference of the knob is divided into 100 equal divisions, so that one revolution will move the scale a specific distance.

The micrometer eyepiece scale is also divided into equal divisions (see Fig. 33, *bottom*) and must be precalibrated with a stage micrometer before measurements are made. The outside drum divisions can then be standardized in terms of the known values of the micrometer eyepiece scale. When measuring an object, therefore, the reading is made on the outside drum scale, which is generally equipped with a vernier, rather than by counting the scale divisions on the micrometer disc. It is essential that the hairline be moved only in one direction for accuracy; otherwise, the mechanical inertia when changing direction may introduce errors in measurement.

Depth Measurement with the Stereomicroscope

The stereomicroscope used for depth measurement (see Fig. 34) is a specially designed instrument that includes a pair of 10X wide-field eyepieces, a pair of 2X objectives, and a dial micrometer gauge mounted on the microscope arm. Target scales are attached to the eyepiece to make certain that the focus is at a definite plane in space. The dial gauge mounted on the microscope arm is capable of reading depth or thickness within a 1-in range, with the smallest graduation being 0.001 inch.

PROCEDURE

The specimen is placed on the stage, and the observer initially focuses sharply on one surface of the detail of the object and then sharply on another surface of the detail. The depth reading is recorded

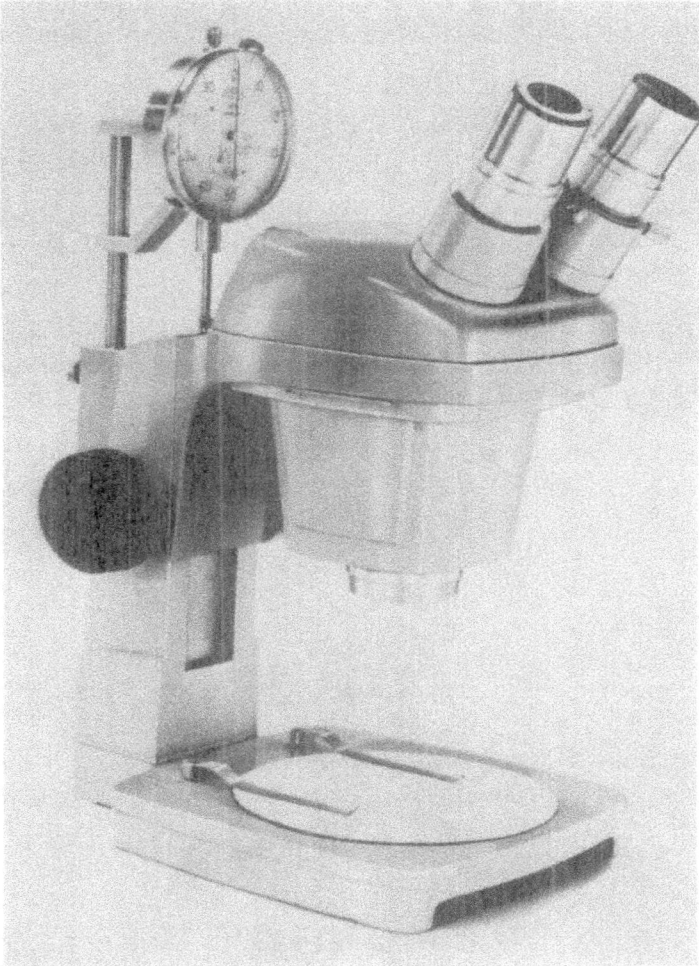

Fig. 34.   Stereomicroscope for depth measurement.

on the dial gauge, which gives the distance traveled from one focus to the other and thus gives the depth measurement.

It is possible to use the standard microscope to obtain an approximate thickness of a specimen by initially focusing, with the fine adjustment, on the upper surface of some detail of the specimen and then on its lower surface. The distance traversed by the fine adjustment is read on the fine-adjustment scale, which is generally graduated in micrometers.

This method is limited, since it is very difficult for the observer to delineate exactly the upper and lower surface of the measured specimen. Moreover, a correction must be made for the refractive index of the mounting medium surrounding the specimen. This can be done by multiplying the observed depth by the refractive index of the medium to obtain the true thickness.

There are also special microscopical instruments that make measurements of depth or thickness. One such instrument, the *optical micrometer* (see Fig. 35), consists of a microscope with a vernier micrometer thimble.

Fig. 35. Optical micrometer.

In use, the objective of this optical instrument is positioned over the specimen. The micrometer thimble is turned until the upper plane of the desired portion of the specimen is brought into sharp focus. A reading is taken on the micrometer scale attachment. The micrometer thimble is then rotated again until the bottom plane of the specimen is brought into sharp focus and a reading is again taken on the micrometer scale. The difference in the two readings will give the depth (that is, the thickness or vertical displacement of the specimen). The actual result must be corrected for the refractive index of the medium surrounding the specimen.

COUNTING WITH THE MICROSCOPE

There are available various counting micrometer discs that are ruled to count specific materials from a known volume. These discs are placed within the eyepiece in a manner similar to that for positioning the measuring micrometer discs.

Howard Disc

The Howard disc (see Fig. 36) is a ruled square subdivided into 36 equal squares; thus, the side of each square is equal to 1/6 the diameter of

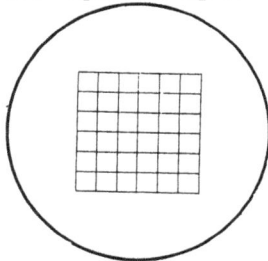

Fig. 36.   Howard disc.

the field.   Since the field of view is limited by the eyepiece diaphragm opening, the side of each square is therefore equal to 1/6 the diameter of the diaphragm opening, generally, of a 10X Huygenian eyepiece. The microscope must be calibrated for the Howard method of counting molds, so that the field of view has a diameter of 1.382 mm at a magnification of 90 to 125X.

The Howard counting disc is used to determine the mold content of food products and to evaluate emulsions and suspensions. A Howard mold-counting chamber is used with it. The chamber consists of a one-piece glass slide with a circular cell (see Fig. 37, *left*)

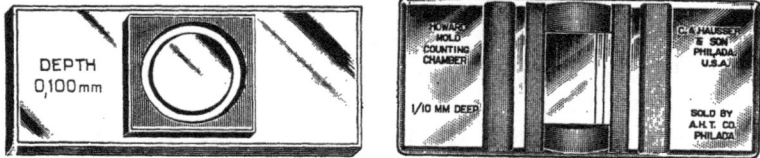

Fig. 37. Howard mold-counting chambers : (left) *circular;* (right) *rectangular.*

19 mm in diameter and 0.1 mm deep in its center, surrounded by a moat or channel to allow for fluid overflow and drainage. Another type of Howard chamber has a rectangular cell (see Fig. 37, *right*), 15 by 20 mm and 0.1 mm deep. Shelves or rests on both sides of the cell support the cover glass. To standardize the microscope to the field diameter at a magnification of 90 to 125x, the slide is engraved with a circle 1.382 mm in diameter, or with two fine parallel lines 1.382 mm apart.

Under the microscope the Howard disc is seen in focus with this special mold-counting chamber. Since the volume of the cell is known, the number of organisms, molds, or particles in a given volume of fluid can be determined.

Sedgwick-Rafter Method

The Whipple micrometer disc is used together with a counting cell in the Sedgwick-Rafter procedure for concentrating, counting, and studying the small plankton found in various bodies of water. The disc can also be used for counting bacteria, dirt particles, and other microorganisms. Its chief use, however, is in the Sedgwick-Rafter method, which is described in detail below.

EQUIPMENT

WHIPPLE MICROMETER DISC

The Whipple micrometer disc, which is placed in the eyepiece of the microscope, is a ruled square subdivided into 100 small squares. The center square is further subdivided into 25 smaller squares (see Fig. 38). The size of the original large square is such that when the

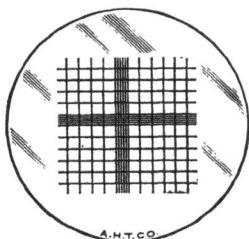

Fig. 38.   Whipple micrometer disc.

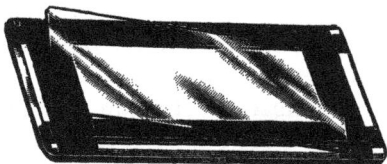

Fig. 39.   Sedgwirck—Rafter Counting Chamber.

Fig. 40.   Cylindrical funnel.

disc is used with a 10X eyepiece and a 10X, 16 mm objective, the large square will cover an area on the stage of exactly 1 mm². Each of the *smallest* squares thus represents an area of 20 by 20 micrometers, or 400 micrometers². This area is known as the *Areal Standard Unit*. The volumetric standard unit has an area of 20 by 20 micrometers and a thickness, or depth, of 20 micrometers. The volumetric standard unit is therefore 8,000 micrometers³.

A stage micrometer must be used so that the dimensions of the squares can be assigned true values. When true values of the squares have been ascertained, the tube length of the microscope is adjusted, if the microscope has an adjustable body tube, so that the area covered on the stage is exactly 1 mm². If the body tube is not adjustable and the area covered by the disc is larger or smaller than the field of view, a calibration factor is used to convert the dimensions of the squares into true values. This is done by standardizing with a stage micrometer (see page 67).

COUNTING CHAMBER

A counting chamber, or cell, with a known area and volume is used in conjunction with the Whipple micrometer disc. The cell consists of a flat optical piece of glass with a centered rectangular brass or glass rim cemented on the glass (see Fig. 39). The depth of the cell is exactly 1 mm, with the inside dimensions being 20 by 50 mm. The capacity of the cell is exactly 1 ml.

A rectangular No. 3 cover glass (0.25 to 0.50 mm thick) of sufficient size is provided to cover and hermetically seal the entire cell when filled.

CYLINDRICAL FUNNEL

A cylindrical funnel (see Fig. 40) is required for filtering the water samples to concentrate the microscopic specimens. The total length of the funnel is about 15 in.; its capacity is 500 ml. The diameter of the top funnel opening is about 2 in.; the sides of the funnel extend straight down for about 10 in. and then narrow for about 3 in. to form a bore at the bottom of the funnel about 0.5 in. in diameter. The funnel can be tightly closed at the bottom by means of a one-hole rubber stopper carrying a small glass U-tube. The arm of the U-tube should extend about 1 in. above the small end of the inserted rubber stopper. The purpose of the U-tube is to prevent complete drainage of water during the filtration procedure.

CLOTH DISCS

Cloth discs, made of 200-mesh silk, bolting cloth, nylon, or linen, and cut in 3/8-in. circles, are required to support the sand during the filtration procedure.

WHITE FILTERING SAND

White sand, consisting of 99.8% pure silica, washed and screened to approximately 60 to 120 mesh, is required for the filtration of the water samples. Ottawa sand, white beach sand, or ground quartz is satisfactory.

PROCEDURE

The water sample should be filtered quickly after it is collected. Insert the prepared glass U-tube into the large end of a one-hole rubber stopper. Do not extend the tube beyond the small end of the stopper. Moisten the stopper end slightly and cover it with a cloth disc. Insert the cloth-covered stopper into the lower end of the funnel. Support the funnel assembly vertically in a stand.

Pour sufficient dry white sand into funnel so that a layer about 1/2 in. deep is formed on the rubber stopper. Add about 10 ml of distilled water to the funnel to wash down the sand and to eliminate any entrapped air in the sand.   As the water filters through the sand, move the funnel gently from side to side to expedite the removal of any entrapped air.

In a graduate, measure 250 to 1,000 ml of the well-mixed sample. It is essential that the volume of the sample taken should produce at least 10 organisms per field.   Tilt the funnel slightly and pour the water sample slowly down the sides of the funnel so that the layer of sand is undisturbed.   Permit the water to filter through the sand. Gentle suction may be used to accelerate the filtration.   Use a little of the filtrate to wash down the sides of the funnel.   The temperature of the liquid during filtration must be kept uniformly at room temperature, since many organisms are very sensitive to a rise in temperature.

When the water in the funnel has reached the same level as in the outer arm of the U-tube, stop the suction and gently remove the U-tube from the rubber stopper.   Allow the remaining water to filter through the sand.

Place a small beaker nearby.   Hold the funnel in a horizontal position and remove the rubber stopper.   Transfer the funnel to the small beaker, hold vertically, and allow the sand to fall into the beaker.   Wash down the walls of the funnel with 5 to 15 ml of the filtrate into the beaker.   The amount of wash water used, which will depend on the desired concentration of the final sample, should be exactly measured with a pipet.

Shake the beaker gently to detach the organisms from the sand. Allow a few seconds for the sand to settle and decant the liquid into another beaker.   Add an additional 5 ml of wash water to the sand, shake gently, allow to settle, and again decant the water into the second beaker. If desired, the concentrated sample may be preserved for future use by adding a 5% formalin solution.

Prepare the cell by placing the cover glass obliquely across it so as to allow the two opposite ends to be uncovered.   Shake and mix the sample gently and, while the water is still agitated, immediately withdraw with a pipet an aliquot 1-ml portion from the bottom of the beaker.   Transfer the sample to the cell, introducing about half the sample into one corner of the cell and the other half into the other corner. The cover glass can now be rotated to cover the cell completely. Allow the organisms to settle for about 5 minutes. Most of the organisms will sink to the bottom; a few will rise to the surface.

COUNTING THE PLANKTON

The American Public Health Association specifies certain rules for counting the plankton in water. The differential count of organisms is the one generally pursued in the Sedgwick-Rafter cell method. The differential count determines the type of organisms present, the identification of these organisms, and the total number of each. In contrast with this procedure, the total-count method determines the total number of plankton present without differentiation of the types.

There are three separate procedures employed in making a differential count of the plankton present in the sample—the field count, the strip count, and the survey count. All three procedures are described below.

The following precautions should be observed when employing any of the three procedures:

1. Count any organisms that are alive or that were alive prior to the addition of a preservative to the sample.

2. Normally, do not count the bacteria present.

3. If an organism is positioned on the boundary line of a field, count the portion of the organism that is inside the field. Consider it as a fraction of an organism.

4. Record the presence of wood fibers, rocks, or other inorganic objects so as to give some information concerning the origin of the sample.

5. If the plankton can barely be seen under 100X magnification, use a higher-power objective.

6. Whenever possible, record each individual organism, unless the specimens are clumped together as colonies or appear in filament form.

7. Clumps or colonies of organisms are recorded in terms of estimated arbitrary units of arbitrary volume.

8. Filamentous organisms are recorded in units of arbitrary length. A standard filament length of 100 micrometers may be considered as one unit.

Field Count

The Whipple micrometer disc is inserted in the 10X eyepiece of the microscope in conjunction with a 10X objective. The counting cell is placed on the stage of the microscope and the Whipple disc is brought into focus with the counting cell. If the organisms are plentiful, only a few fields need be counted for the determination of the plankton count. The actual number of fields to be counted to obtain a good mathematical distribution and accurate count will depend on the worker. For a high degree of accuracy or where special research studies are made, at least 25 fields or more should be counted.

For routine work, where a field contains a minimum of 10 organisms, 10 standard fields will be sufficient to give an accurate count. If a 40X objective is used in lieu of the 10X objective, the number of fields routinely examined should be at least 40. The fields examined should be well separated from each other and should be chosen at random from different areas of the cell.

The dimensions of the counting cell are 20 by 50 mm by 1 mm deep, giving a volume of 1 ml. The area covered by the Whipple disc is 1 mm². Thus, each field covers 1/1,000 ml of the aliquot sample or concentrate and 20 fields will cover 1/50 ml.

After the examination, record the total count, the species present, and the number of each type of organism seen.

Strip Count

This method is used when certain organisms do not appear in satisfactory numbers in each field examined. The strip count is made when there is less than one selected group of organisms per observed field and when an accurate count cannot be made with the Whipple disc. This method is also employed when the observed organisms are too small to be examined and counted properly under 100X magnification. The method also endeavors to obtain the count of a select species of organisms as they occur in the full length of the counting cell when using a 10X eyepiece and a 40X, 8 mm objective. The dimensions of the microscope field will then be 50 by about 0.7 mm. The actual width may be determined by using a stage micrometer.

Begin counting the organisms at one end of the cell. Continue counting as the slide is slowly moved by the mechanical stage along the long axis of the cell and past the objective until the other end of the cell is reached. Usually one traverse or trip is made, although for greater accuracy several strips may be counted and the average taken.

If a cell 20 by 50 mm by 1 mm deep is used and the Whipple disc covers a field 1 mm square, then each traverse will cover 1/20 ml of the aliquot sample. This calculation presupposes 100X magnification.

Survey Count

Large organisms, such as rotifers, worms, or microcrustaceans that appear less numerous than the smaller organisms in the sample, are counted by the survey method. Since these organisms are rather large and easily visible under low magnification, a stereomicroscope is employed to count the total number of a specific type of organism in the entire cell volume of 1 ml.

An optional procedure with the stereomicroscope, if the plankton is too small, is to transfer the entire contents of the cell into a Petri dish for examination. To expedite the counting procedure, diagonals or squares are ruled on the dish. If a stereomicroscope is not available, the standard microscope, using a 10X eyepiece and a 10X objective, can be utilized to count the plankton in the counting cell.

The counting cell is placed in the mechanical stage, and the entire cell is moved slowly past the objective. A record of the count of the total number of organisms is kept.

'Milk Smear' Micrometer Disc

The "milk smear" micrometer disc is used to count the number of bacteria when a definite volume of milk is spread over a known area on a glass slide. The micrometer disc has ruled on it an 8.0-mm circle, divided into quadrants to facilitate counting. The accuracy of the count depends on an accurate determination, by means of a stage micrometer, of the area on the slide covered by the circle. The disc is placed in the 10X eyepiece so that it rests with the etched surface downward on the diaphragm. An oil-immersion objective is used with the 10X eyepiece.

The area of the microscope field must be standardized or calibrated so that an accurate bacteria count can be achieved. It has been estimated that a field diameter of 0.206 mm will give a *microscope factor* of 300,000. Microscope factors are used to determine the average number of bacterial clumps or the average number of individual bacteria per milliliter of milk sample. For example, the total number of bacteria counted in one field multiplied by the factor 300,000 will give the total number of bacteria per milliliter of milk. If 50 separate fields are used to determine the bacteria count, the total number of bacteria in 50 fields multiplied by 6,000 will give the total number of bacteria per milliliter of milk. The number of fields counted will depend on the accuracy desired. Generally, 30 fields will give a sufficiently accurate count.

A 10X ocular in conjunction with an oil-immersion objective (1.8 mm) can be adjusted to give a field diameter of 0.160 mm and a microscopic factor of 500,000. Field diameters of 0.178 and 0.146 mm will give microscope factors of 400,000 and 600,000, respectively.

The calibration of the microscope to give the desired field diameter can be accomplished by reducing the aperture of the diaphragm found within the eyepiece. This reduction can be accomplished by placing on the diaphragm platform a circle of cardboard with a hole of specific diameter. If the field is still too large, other cardboard circles, with smaller holes, are prepared so that one will have the correct measurement.

Special eyepiece discs are designed with concentric inscribed circles for adjusting the field diameter. This is done by choosing the inscribed circle that encloses the exact field of view. It must be observed that once a calibration is made, the same combination of objective, eyepiece, and tube length must be used to obtain a similar microscope factor.

PROCEDURE FOR CALIBRATING THE MICROSCOPE

Place a stage micrometer on the stage. Place the milk-smear

micrometer disc with the reducing diaphragm on the platform within the eyepiece. Bring the immersion objective in line and measure the field diameter in millimeters to the third decimal (as, for instance, 0.160).

PROCEDURE FOR DOING BACTERIA COUNTS ON MILK AND CREAM

Shake the sample well. Insert a clean 0.01-ml capillary pipet into the sample and draw up the exact portion. Place the tip of the pipet on a glass slide and carefully discharge the milk sample onto the slide. Spread the sample evenly over 1 cm, using a sterile needle. Place the slide on a flat surface and dry quickly at 40 to 45°C. Drying time should be no more than 5 min. During the drying procedure, the slide must be protected from dust or any contamination.

Immerse the dry smear in xylol for 1 to 2 min. to extract or remove the fat. Remove the slide, dry it, and immerse it in ethyl alcohol (95%) for 1 to 2 min. The slide is now ready for staining. One of three prepared stains may be employed, the choice depending on the worker. (See Appendix for the preparation of these special stains.)

METHYLENE BLUE STAIN

Immerse the slide vertically into the stain solution for about 2 min. Lower and raise the slide several times to detach any air bubbles. Remove the slide from the stain solution and allow to drain. Rinse immediately in water until the excess stain is removed. Dry in air before examining under the microscope.

If the slide is too deeply stained, decolorize in ethyl alcohol (95%) for about 20 to 50 sec. A good slide will be colored pale blue. Slides can be restained and decolorized as many times as necessary without injury until the right degree of stain is achieved. Apply cedar oil to the slide and examine with the oil-immersion lens.

NEWMAN-LAMPERT STAIN (MODIFIED BY LEVOWITZ AND WEBER)

Fix the slide as described above. Immerse the fixed slide vertically into the stain for about 2 min. Remove the slide from the stain solution and allow to drain by placing the slide on absorbent paper. Dry slide thoroughly in air. Drying can be hastened by gently blowing air over the slide. Rinse the dried slide three times in tepid water (100 to 110°F) and again dry thoroughly in air.   Apply cedar oil to the slide and examine under the oil-immersion objective.

NORTH'S METHYLENE BLUE-ANILINE OIL STAIN

Fix the slide as described under methylene blue stain.   Immerse the fixed slide in the stain for 1 min.   Remove the excess stain by dipping the slide several times in a vessel of fresh water.   Allow the slide to drain and then dry thoroughly in air.   Apply cedar oil to the slide and examine under an oil-immersion objective.

COUNTING THE BACTERIA

When observing the milk slide under the oil-immersion objective, incomplete divided bacteria are counted as two; a clump is regarded as one bacteria as long as not too many are congregated together and bacteria that are within cells are included in the count.   This microscopical method of counting bacteria in milk has a practical application in grading or setting up standards for milk.   For Grade A milk that is to be pasteurized, the slide count must not exceed 200,000 per milliliter of milk.   For milk that is to be consumed raw, the count must not exceed 50,000.   For Grade A cream, these counts are doubled.

Net Micrometer Disc

The net micrometer disc (see Fig. 41) is ruled with a 5-mm square subdivided into 100 equal squares.   It is used for counting large particles.   The disc must be standardized with a stage micrometer before use.

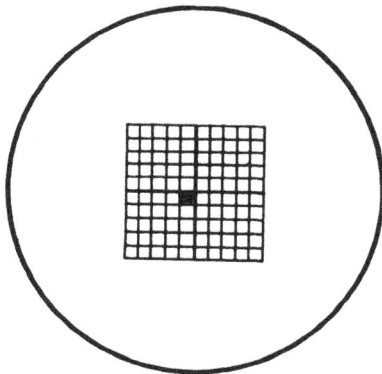

Fig. 41.   Net micrometer disc.

Hemacytometer

The hemacytometer is another type of counting chamber that also incorporates accurately ruled squares on its surface so that no ocular micrometer need be used. It is a precise counting chamber and finds its chief use in red and white blood-cell counts, platelet counts and spinal-fluid counts. It can also be employed to count various small particles that may be in suspension—microorganisms, yeast, and dust.

Several types of hemacytometers are available. These may differ in the design of the chambers and rulings, but they are all used in a similar manner. The chamber usually consists of a thick glass slide with two platforms, centrally located and facing each other; surrounding the two platforms is an H-shaped moat (see Fig. 42, *top*). These platforms are engraved with a series of rulings and are flanked by two other platforms called the *slide platforms*, which extend exactly 0.1 mm above the ruled platforms. A special optical-plane cover glass, measuring 20 by 26 mm and from 0.4 to 0.6 mm thick, fits exactly over the two raised platforms. The ruled surfaces will therefore be 0.1 mm below the cover glass, with each ruled plateau having a surface area of 9 mm^2.

The ruling of the hemacytometer scale is illustrated in Fig. 42, *bottom*. The central square, with an area of 1 mm^2, is divided by triple lines into 25 smaller squares, each of which contains a group of 16 still smaller squares, for a total of 400 squares, each with a definite boundary and measuring 1/400 mm^2 in area.

The central square area, generally used for the red-cell count, limits the volume of fluid to 0.1 mm^3, since the area of the central square is 1 mm^2 and the depth from the cover glass to the ruled surface is 0.1 mm. The volume over each of the 400 subdivided squares will be 0.00025 mm^3.

Located outside the triple ruled lines are four corner squares (1 mm^2), each subdivided into 16 smaller squares. The four corner squares, together with the central square, are used to count white blood cells.

Special pipets are used with the hemacytometer to dilute the blood specimens. The Thoma Red diluting pipet (see Fig. 43, *top*) used for

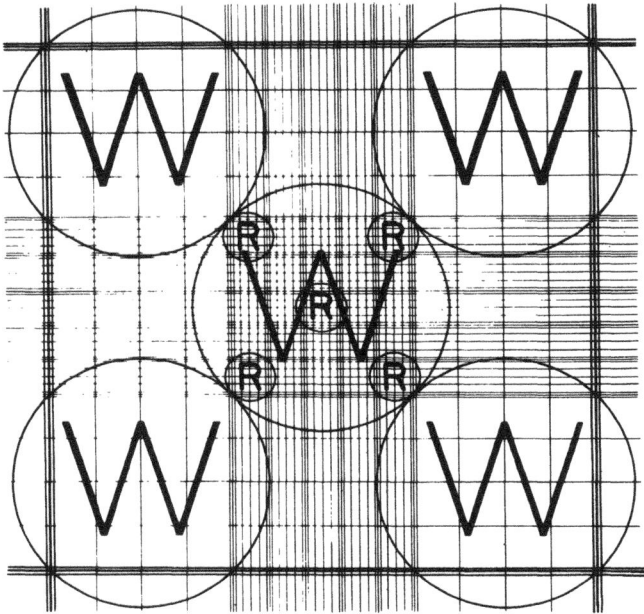

Fig. 42. Hemacytometer (top) *and hemacytometer scale* (bottom); (R) *squares used for counting red blood cells;* (W) *squares used for counting white blood cells.*

counting red blood cells, is graduated to dilute the specimen in the ratio of 1:100 and 1:200, and is identified by a small red bead in the bulb that is also of help in mixing the diluted blood sample. The pipet is graduated along its capillary bore in tenths. The fifth graduation from the tip is marked *0.5* and the tenth graduation is marked *1*. A

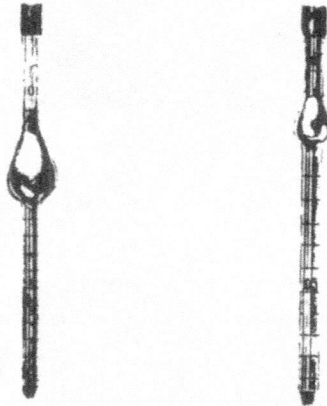

Fig. 43. Red Thoma pipet (left) *and White Thoma pipet* (right).

line etched above the bulb is marked *101*. If the blood is drawn to the 0.5 mark and then diluted to 101, the dilution will be 1:200. If the blood is drawn to 1 and then diluted to 101, the dilution will be 1:100. A red mouthpiece connected to a 10-in. rubber suction tube is supplied with the pipet to aid in obtaining and diluting the blood sample.

The Thoma White diluting pipet (see Fig. 43, *bottom*), used for counting white blood cells, is similar to the Red, the differences being that the bulb is smaller and the pipet is graduated for dilutions in the ratio of 1:10 and 1:20. It contains a small white clear bead in the bulb for quick identification and for aid in mixing the diluted blood. It is graduated along its capillary bore in tenths. The fifth graduation from the tip is marked *0.5*, the tenth graduation *1*. A line etched above the bulb is marked *11*. If the blood is drawn to 0.5 and then diluted to 11, the dilution will be 1:20. If the blood is drawn to 1 and then diluted to 11, the dilution will be 1:10. A white mouthpiece connected to a 10-in. rubber suction tube is supplied with the pipet to aid in obtaining and diluting the blood sample.

The equipment necessary for the red-cell count is the same as for the white-cell count, except for the type of pipet and the diluting solution. For red-cell counts, a diluting solution called Hayem's Solution is prepared according to the following formula :

Sodium sulfate	5.0 g
Sodium chloride	1.0 g
Mercuric chloride	0.5 g
Distilled water	200 ml

The diluting solution for white-cell (leucocyte) counts is a mixture of glacial acetic acid (3 ml) and distilled water (97 ml), with a drop of gentian violet (1%) to color it for convenient identification.

The procedure for obtaining and calculating the red-cell count in a blood sample is described below.

CLEANING THE GLASSWARE

The pipets and counting chamber must be thoroughly clean and dry before use. When cleaning the hemacytometer, do not use harsh abrasives or strong detergents, since they may impair the volumetric accuracy of the chamber. Generally, a mild soap solution and a thorough rinsing will be adequate. When handling the clean, dry chamber, hold it at the sides to eliminate grease stains from the fingers. Better yet, handle the chamber with metal tongs. The pipets are cleaned and rinsed with water by means of suction. They are then dried by drawing alcohol through to remove the water and then acetone or ether is used to remove the alcohol. The acetone or ether is then removed by drawing dry, clean air through the pipets for a few seconds. If the blood in the pipet moves freely when it is shaken, it can be assumed that the pipet was clean and dry.

PROCEDURE FOR RED-CELL (ERYTHROCYTE) COUNTS

Position the cover glass on the hemacytometer. Using the mouthpiece and rubber tube attached to the Thoma Red pipet, draw up the blood specimen exactly to the 0.5 mark. Remove the excess drop from the tip of the pipet with a piece of gauze. Draw up

Hayem's Solution exactly to the *101* mark. This will dilute the blood to 1 : 200.

Holding the pipet in one hand, place the tip against the ball of the thumb of the other to prevent any loss of liquid. Kink the rubber tube at the end of the pipet and hold this end against the middle finger of the same hand. To ensure adequate mixing of the solution, shake the pipet gently for about 2 min.

Place the mouthpiece back into the mouth to control and regulate the flow. Discard at least 4 drops from the pipet, touch the tip of the pipet to the edge of the cover glass positioned on the platform of the counting chamber, and allow a thin layer of fluid to flow under the cover glass. If any air bubbles appear under the cover glass, or if the liquid flows over into the H-moat, clean and dry the chamber and repeat the procedure.

Permit the red cells to settle for 2 min. Place the counting chamber on the stage of the microscope. Use a 10X eyepiece and a 10X objective for the examination of the chamber. Bring the chamber into focus and note whether the red cells are evenly distributed. If not, clean and dry the chamber, shake the pipet again, and flow another sample under the cover glass.

COUNTING THE RED CELLS

The counting of the red cells is done in the central large square (see Fig. 44). Count all the red cells in each of the corner squares and in the center square enclosed by the triple lines. It can be seen that each of these five counting squares is composed of 16 smaller squares that lie within the triple lines. This procedure is followed to obtain an average estimated count if the blood sample is evenly distributed in the chamber.

Count the cells within each of the five squares mentioned above. Also, count any cells touching the triple lines on the top and right of the square. Disregard any cells touching the triple lines on the bottom and left of the square. The number of red cells per cubic centimeter can then be calculated from the total count of the five squares. If there is an excessively uneven distribution of cells, repeat the preparation of the blood sample and its count.

CALCULATING THE RED-CELL COUNT

To simplify the explanation of the procedure for calculating the red-cell count, a hypothetical count is included here.

1.   Assume the counts in the five squares to be 85, 104, 103, 126, and 92 red cells, for a total of 510.

2.   The total of smaller squares within the five counting squares is $16 \times 5 = 80$. The total of small squares in the main central large square (1 mm²) is 400.   Therefore, 80/400, or 1/5, mm² is the area occupied by the five counting squares.

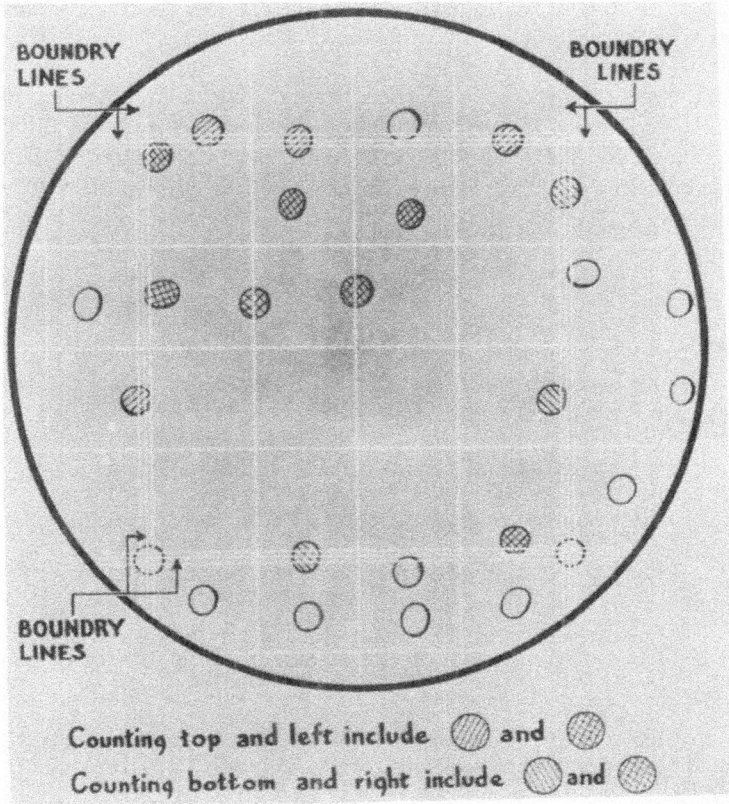

Fig. 44.   Counting red and white blood cells (discussed in the text).

3. The number of red cells within an area of 1 mm^2 would then be $510 \times 5 = 2{,}550$ red cells.

4. The cell depth under the cover glass of the counting chamber is 0.1 mm. Hence, the number of red cells of the diluted blood specimen per cubic millimeter would be $10 \times 2{,}550 = 25{,}500$.

5. The blood was diluted 1:200. Therefore, the actual red-cell count of the undiluted sample is $200 \times 25{,}500 = 5{,}100{,}000$ red cells per mm^3.

6. The calculation can be summarized in the equation: red cells per cubic millimeter of blood = cell count (in five counting squares) $\times 5$ (area) $\times 10$ (volume) $\times 200$ (dilution).

A quick method can be used for calculating the blood count per cubic millimeter. If five counting squares are used and the dilution of the blood is 1:200, add four zeros to the total count derived from the five counting squares. For example, if this count is 510, the blood count is 5,100,000/mm^3.

The normal red-cell (erythrocyte) count for males ranges from 4,600,000 to 6,200,000; for females, from 4,200,000 to 5,400,000.

PRECAUTIONS

The following precautions must be observed in using the hemacytometer for red-cell counts:

1. A good sample of blood must be initially obtained.

2. The chamber and pipets must be thoroughly clean.

3. The sample of blood must be drawn up the pipet exactly to the mark.

4. The excess of blood on the outside of the pipet must be wiped off.

5. The diluting solution must be fresh and uncontaminated and it must be drawn up the pipet exactly to the 101 mark.

6. The pipet must be shaken sufficiently so that the cells are evenly distributed in the solution.

7. The diluted blood sample must not remain in the pipet for too long a period, since the cells will settle.

8. The counting chamber must not be overfilled.

9. There should be an even distribution of the sample in the chamber.

10. Sufficient time must be allowed for the red cells to settle in the chamber.

11. Visual errors in counting the cells must be avoided, since otherwise low counts may result.

12. If greater accuracy is desired, more squares must be counted.

PROCEDURE FOR WHITE-CELL (LEUCOCYTE) COUNTS

The general procedure for white-cell counts is similar to the one described for red-cell counts.

First draw the blood specimen to the *0.5* mark in the Thoma White pipet and then draw the diluting fluid (special for white-cell counts) to the *11* mark.   This makes a dilution of 1:20.

Shake the pipet in the same way as for the red-cell counts, discard at least 4 drops, fill the counting chamber, and allow the cells to settle.

Examine the chamber, using a 10X eyepiece and a 10X objective.

COUNTING THE WHITE CELLS

.Count the white cells in each of the four large corner squares. Each of these counting squares is subdivided into 16 smaller squares. The difference in count between the largest and smallest number of white cells found in any two of the counting squares should not exceed 10 cells. For greater accuracy, the large central square may be included in the count.

CALCULATING THE WHITE-CELL COUNT

To simplify the understanding of the procedure for calculating the white-cell count a hypothetical count will be given.

1. Assume the counts in the four corner counting squares to be 32, 40, 36, and 30, for a total of 138 white cells.

2. The dimension of each counting square is 1 mm², and each contains an average of 138/4, or 34.5, white cells.

3. The cell depth is 0.1 mm. Therefore, 1 mm³ contains 34.5 × 10, or 345, white cells.

4. The blood was diluted 1:20. The white cell count is 20 × 345, or 6,900 per mm³.

5. A formula for computing the white-cell count is : white cells per cubic millimeter=average count of one counting square × 10 (volume) × 20 (dilution).

A short method for calculating the white-cell count would be to multiply the total number of cells found in four large counting squares by 50. For example, the number found in four counting squares is 138. The total white cells per cubic millimeter would be 138 × 50 or 6,900. The white-cell count for normal blood ranges from 5,000 to 10,000.

PRECAUTIONS

The precautions listed under the procedure for the red-cell count should also be observed for the white. In addition, the following facts should be known :

1 The leucocyte count in normal blood is variable with the time of day. If a daily count is taken, it would therefore be taken at approximately the same time each day.

2. When white-cell counts are very high, a dilution of 1:100 may be necessary. In such a case, use the Thoma Red pipet to dilute the blood and change the calculation accordingly.

3. If white-cell counts are extremely low, draw the blood to the *1* mark in the white-cell pipet, which makes the dilution of the blood 1:10. To obtain the total white-cell count, the calculation is altered by multiplying the total number of white cells counted in four large counting squares by 25.

Using a Reticulocyte Micrometer Disc

If cresyl blue is mixed with the fresh red cells of a blood sample, some of the cells will give the appearance of a blue reticulogranuler network when observed under the microscope with the oil-immersion objective. These cells, called *reticulocytes,* give an approximation of the rate of blood corpuscle production. Normal blood contains approximately 5 to 10 reticulocytes per 1,000 red cells.

The reticulocyte micrometer disc (see Fig. 45) is used to facilitate the counting of reticulocytes. Engraved on the disc surface are two

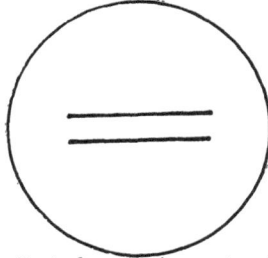

Fig. 45.   Reticulocyte micrometer disc.

parallel lines 2 mm apart. The disc is placed on the diaphragm within the 10X Huygenian eyepiece and is used in conjunction with an oil-immersion objective.

PROCEDURE FOR THE WET-FILM METHOD

Step 1.   Place a small drop of blood in the center of a cover glass. Add two small drops of the cresyl blue stain (1.0 of brilliant cresyl blue and 0.4 g of sodium citrate dissolved in 100 ml of physiologic saline solution).

Step 2.   Invert the cover glass on a slide. Remove the excess liquid by touching the edges of the cover glass with a piece of gauze. A thin preparation will give a better and easier count.

Step 3.   Surround the cover glass with vaseline to prevent evaporation and allow to stand for 6 to 10 min.

Step 4.   Place the slide on the microscope stage and focus the specimen by using the oil-immersion objective.

Step 5.   Count 1,000 red blood cells—200 in the center and 200 in each of the four corners within the two parallel lines limited by the disc.

During the count, note the number of cells that have bluish granules or a blue reticular network. The number of such cells divided by 10 will give the percentage of reticulocytes present in the blood.

CHAPTER 5

Preparing Mounts for the Microscope

THE METHODS OF MOUNTING various transparent or translucent specimens can be generally classified as wet mounting, dry mounting, and balsam mounting. The necessary materials for preparing and mounting the specimens will be described with each procedure. It is essential that these materials be kept scrupulously clean, free of all dirt and grease.

SELECTING THE GLASS SLIDE

A thoroughly cleaned glass slide, measuring 75 by 25 mm and 1.0 to 1.25 mm thick, is commonly used for mounting specimens. A good glass slide, when purchased, will be precleaned, plane-surfaced, of uniform thickness, and with a polished surface, rounded corners, and ground edges. It will have good optical qualities and be free from scratches and bubbles. It will be resistant to surface attack from chemical reagents and weathering. Glass slides bought from reliable companies will be fabricated to these specifications and will be acceptable for most microscopic work.

Standard slides are available either plain (see Fig. 46) or with a frosted strip on one or both ends (see Fig. 47). The frosted strip which takes lead pencil, crayon, or ink, is convenient for numbering or other identification of the mount.

*Fig. 46.   Plain
glass slides.*

Special glass slides for microcultures are designed with one or two depressions or concavities about 0.8 mm deep ground into its top

Fig. 47.   Glass slide with frosted strip on one end.

surface (see Fig. 48).   In use, a cover glass is placed over the cavity to provide a seal that protects the culture preparation from drying out.

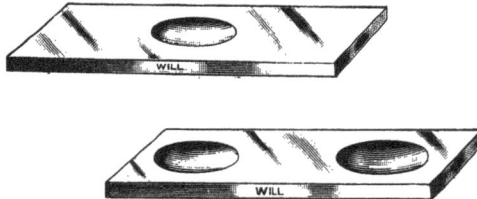

Fig. 48.   Glass slides with concavities.

Thick glass slides with one or two deeper concavities 1.75 to 3 mm deep are also commercially available.   These are suitable for the critical examination of drop cultures, agglutination tests, medical research for tissue, protozoan culture study, and many other uses.

SELECTING THE COVER GLASS

When mounted on a slide, the specimen is generally covered with a cover glass (see Fig. 49) for protection. Since the glass slide and the cover glass are both part of the optical system, it is essential that both

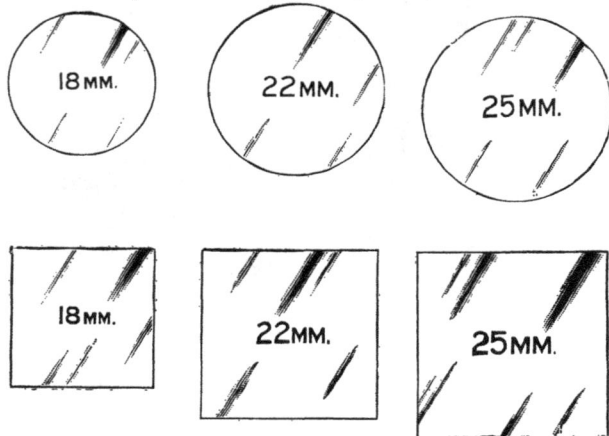

Fig. 49. *Round and square cover glasses (shown actual size).*

be of the best quality obtainable in order to allow the best possible image of the specimen to be achieved.

The cover glass should be flat, nonfogging, and free of bubbles, scratches, and striations. It must be scrupulously clean and 0.18 ± 0.003 mm thick. The ideal thickness range is 0.17–0.18 mm.

Good objectives (dry) are corrected for spherical aberration and generally give the best image with a cover glass of the proper thickness (0.17–0.18 mm). A cover glass that is too thick may overcorrect for spherical aberration, while one that is too thin may undercorrect, thus causing a loss in quality of the observed image. It is therefore essential that the correct thickness of cover glass be used with dry objectives, especially when photomicrographs are to be taken.

The thickness of the cover glass will not be too critical when used with an oil-immersion objective, since the refractive index of the immersion oil will be almost equal to that of the cover glass. For best results, however, it is essential that a cover glass of suitable thickness range be used.

Cover glasses are available in square, rectangular, and round shapes. They are also available in different thicknesses, as follows:

No.	Thickness (mm)
00	0.05–0.085
0	0.085–0.13
1	0.13–0.16
1½	0.16–0.19
2	0.19–0.25
3	0.25–0.35

Available sizes of square and round cover glasses are 12, 15, 18, 22, and 25 mm. The rectangular cover slips are generally used for special purposes such as blood smears. Round covers are preferred for most uses, since it is much easier to "ring" the cover to seal the mount or to give the slide a neater appearance.

Cover glasses are sold by weight (unit package weight, one oz.); the dimensions of the cover glass chosen should be sufficient to cover both the specimen and some additional surrounding area.

CLEANING GLASS SLIDES AND COVER GLASSES

Many of the slides and covers received from a supplier may not be absolutely clean, so it is good practice to clean all slides and cover glasses and store them in a dust-tight container. Cleaning can be done by immersing them in a hot soap or detergent solution for several minutes and then rinsing them thoroughly in running water. If necessary, the surfaces of the slides can be scrubbed with a small, stiff brush. To hasten drying, after they are rinsed in water, they can be immersed in alcohol. Then they should be dried and polished with a soft, lint-free linen cloth. Handle the slides by their edges, without touching the flat surfaces, rub them until they are polished and then hold them up to the light for inspection.

Since cover glasses are thin and fragile, they must be handled more gently than slides. Place the cover glass in the fold of the cloth as in a sandwich. Grasp the cover glass and cloth between the thumb and forefinger of one hand and then move the cloth gently, using a circular motion, across each side of the cover glass, exerting equal pressure on each side so that the cover can be dried and polished without breaking. Another method of polishing cover glasses is to place them between two cloth-covered blocks of wood and polish them with a gentle rotary motion.

Clean slides should be stored in a clean, dry receptacle, covered to exclude dirt. They can also be stored in a special microslide dispenser, which can hold about 50 to 60 slides. These dispensers are equipped with a lever or handle to eject slides singly when needed. When these dispensers are not available, the slides are stored in a covered glass receptacle. One type of receptacle is the high-walled Stender dish, which has grooved sides to keep the slides separated. The Stender dish is used primarily for preparing microscopic specimens and for fixing, dehydrating, and staining tissues. *Do not store slides by piling them one on top of the other.*

Low, round Stender dishes (see Fig. 50) are useful for storing cover glasses. These dishes have ground edges and covers of glass ground

Fig. 50. Stender Dishes.

to fit the dish so as to exclude dirt and moisture. Always handle clean cover glasses with forceps or hold them by the edges to avoid touching the surfaces.

PREPARING A WET MOUNT

The term wet mount denotes a specimen surrounded by liquid medium of fluid or gelatinous consistency. Fluid mounts are used for living specimens in their natural environment or, if dead, they will appear less distorted and give a more natural appearance. There are various liquids that can be used for fluid mounting—among them are water, glycerin, neutral salt solution, and formalin (3%). The choice of fluid will depend on what is necessary for correct preparation of the specimen.

Procedure

With a medicine dropper or a pipet held in the hand, submerge the tip in the culture at the point where the specimens are to be taken (samples should be taken from the bottom, middle, and top layers of each culture to assure a good sampling of organisms).

Place one or two drops of the culture in the center of the glass slide. Always hold the slide and cover glass by their edges to prevent smudging. Allow the cover glass to fall gently on the specimen in such a way as to prevent any entrapment of air bubbles. A good procedure is to hold the cover glass by the fingers along the edges—or, better still, with forceps—at approximately a 45° slant. The edge of the cover glass should touch the slide just beyond the culture drop. Lower the cover glass gently until it covers the specimen and forces the liquid to spread out into a thin film without entrapped air bubbles. This gentle lowering of the cover glass over the specimen can be done with the aid of a dissecting needle.

Let the slide lie undisturbed for about two minutes before observing the organisms under the microscope. If an excess of liquid oozes beyond the edges of the cover glass, prepare another mount using a slightly smaller quantity of the liquid culture.

If the specimen is to be observed for a long time, the liquid under the cover glass may evaporate. On the average, temporary aqueous mounts will last about a half hour, after which evaporation of the water will draw the cover glass closer to the slide, compressing and injuring any live organisms present. In those cases where evaporation hinders the study of the live organisms, the evaporated liquid must be replaced about every 15 minutes. This is done by placing the mouth of the pipet or medicine dropper containing the original culture exactly at the edge of the cover glass and easing a small drop from the pipet or dropper to the slide; capillary attraction will cause the added liquid to flow under the cover glass and unite with the original liquid. Care must be taken so that none of the liquid flows on top of the cover glass. If it is desired to remove the original liquid and to replace it with new liquid, use the following procedure : Add drops of the replacement fluid at one side of the cover glass as described while simultaneously touching the other side of the cover glass with an absorbent paper such as filter paper. The absorbent paper will draw off the old liquid.

STAINING LIVE SPECIMENS

To stain live specimens or temporary wet mounts, place a drop of the prepared stain at the margin of the cover glass. The stain will then join the main body of liquid under the cover and stain any organism present. When applying stains to live organisms, it is essential that the stains not be injurious or toxic to the specimens.

SEALING A WET MOUNT

If the slide is to be observed over a long interval of time and additional liquid will not be added to replace the evaporated fluid, it will then be necessary to seal the cover glass on the glass slide to prevent any evaporation. This may be done simply by spreading a thin film of petrolatum around the edge of the cover slip with the finger. The petrolatum should spread over the edge of the cover glass onto the glass slide to make an airtight seal.

Another method of preventing evaporation of the liquid is to prepare a cover glass in advance by applying a ring of petrolatum around the edge of the cover glass and then inverting the cover glass over the liquid specimen. For sealing preparations, a circular cover glass is better than a square or rectangular one.

DECREASING MOBILITY OF THE SPECIMEN

It may sometimes be necessary to slow down rapidly moving organisms so that they may be more satisfactorily observed and studied. Several methods of doing this are described below.

ENTRAPMENT WITH A MAT

A thin mat of cotton fibers or a piece of lens paper is placed on the slide. The culture is placed on the mat and covered with a cover glass. The fibers of the mat entangle and slow down the moving organisms. This procedure, however, will often interfere with microscopic observation.

USING A VISCOUS MEDIUM

Add a drop of a viscous solution made from carboxymethyl cellulose (0.1%), gelatin, gum arabic, or other gums. The consistency of the solution necessary to slow down and quiet the specimen without distortion can be determined by trial and error.

ANESTHETIZING THE ORGANISMS

Another extensively used method of slowing down organisms is to add a drop of a narcotic or poison to the culture. For example, one drop of formalin in 100 ml of the culture will slow down the organisms, without killing them, for about 10 minutes. Eventually, however, the organisms will become distorted and die. Narcotics such as ether, chloroform, chloral hydrate, magnesium chloride (1%), nicotine (0.1%), and barbiturates (1:1000 dilution) have also been used for this purpose.

EVAPORATING THE MEDIUM

Allow the liquid on the slide to dry by evaporation. As the liquid dries, the cover glass is drawn closer to the slide and, simultaneously, to the organisms. Prior to complete evaporation, the organisms are greatly restricted in their movements. It should be noted, however, that this procedure may produce some distortion in structure of the organisms.

Wet Mount—'Hanging Drop' Method

It is very advantageous to study living microorganisms by the "hanging drop" technique. This method is used primarily to determine whether an organism is mobile. The equipment required is a special thick slide with a well or cavity in the center, a cover glass, and, if the preparation is to be examined over a long period of time, some petrolatum.

PROCEDURE

Place a thin cover glass (No. 1) on a flat surface. Spread a thin layer of petrolatum around the rim, leaving a clear space in the center of the cover. Place a drop or loopful of the material to be examined in the center of the cover glass. The drop should be sufficiently large so that it will remain suspended from the cover glass without touching any part of the glass slide. Place a glass slide upside down over the cover glass so that the well in the slide is directly over the drop. Press the slide down firmly enough so that the cover glass sticks to the slide. Invert the slide and cover glass quickly, so that the drop culture does not slide and adhere to the wall of the well. The drop will now hang from the cover glass, and the organisms will be free to move within the drop without being pressed down by the cover glass (see Fig. 51).

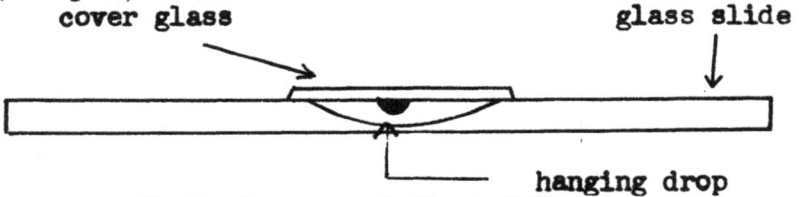

Fig. 51. Cross section of a "hanging" drop mount.

It is also possible to fill the cavity in the slide with the culture, seal a cover glass over it, and thus allow the organisms to move freely in the medium. Under these conditions, however, the range of movement may not be sufficiently limited.

When attempting to find the organisms in the hanging drop with the microscope, first reduce the transmitted light to a minimum by closing the iris diaphragm; then locate and center the edge of the drop under the low-power objective. The high-power objective is brought into position and the portion of the drop is brought into sharp focus. If necessary, the iris diaphragm can be adjusted to achieve the correct illumination.

Preparing Smears and Films

Smears, which include thin films of a fluid culture, are routinely used for examining specimens, especially for studying the anatomical structures of an organism, its cell morphology, the arrangement of cells in the organism, and the reactions of these cell structures to stains. The procedures for preparing smears of several varieties of cultures are described below. These procedures can generally be followed when preparing smears of allied materials.

BLOOD SMEARS

This procedure is widely used for the examination of blood and liquids containing bacteria.

Glass slides or cover glasses can be used in obtaining blood films or smears. Generally, the cover glass gives a better distribution of white cells, and the slide smear gives a better distribution of red cells. Both procedures are described below.

OBTAINING THE BLOOD SPECIMEN

When obtaining a drop of blood for the preparation of a smear, care must be exercised so that no infection occurs.

The part of the skin to be pricked (usually the tip of the finger or the base of the thumbnail) is first wiped with cloth or cotton saturated with alcohol. Sterilization is achieved by passing the needle through a flame or immersing it in alcohol (95%) or tincture of iodine, and then wiping with a sterile gauze. Individually packaged sterile needles are commercially available.

Prick the finger with the needle, using a quick, short stabbing motion. A drop of blood will gather on the surface of the finger. Wipe off the first drop with a sterile gauze, then squeeze the finger below the puncture so that the second drop of blood will become enlarged. After the sample of blood is obtained, sterilize the puncture by applying a sterile gauze moistened with alcohol.

BLOOD SMEAR WITH COVER GLASSES

Choose two cover glasses (No. 2) 22 mm square that are absolutely flat and clean. Handle the cover glasses only by their edges. Without touching the finger, hold the cover glass directly above the second drop of blood obtained as described above and allow the drop to touch the cover glass (the drop of blood on the cover glass will be small). Place the second cover slip over the one containing the blood sample and spread it so that no air bubbles are introduced into the drop of blood. When the blood has almost stopped spreading, slide the glasses apart along the horizontal axis. Never pull upward, as this will ruin the smear.

Transfer the cover glasses, with the smeared sides up, to a clean countertop and allow to dry at room temperature. A good smear will be without streaks and show few holes when observed under low-power magnification.

BLOOD SMEAR WITH GLASS SLIDES

The technique of preparing a blood smear with two glass slides is illustrated is Fig. 52. Place a small drop of blood about an inch from the end of a glass slide. Another method is to touch the pricked finger with one end of the glass slide. Place the glass slide on a flat surface. Place one end of another glass slide in front of the blood drop, keeping this slide inclined at about 30°, with the greater part of this second slide not over the specimen slide or coincident with it but rather behind it (see Fig. 52).

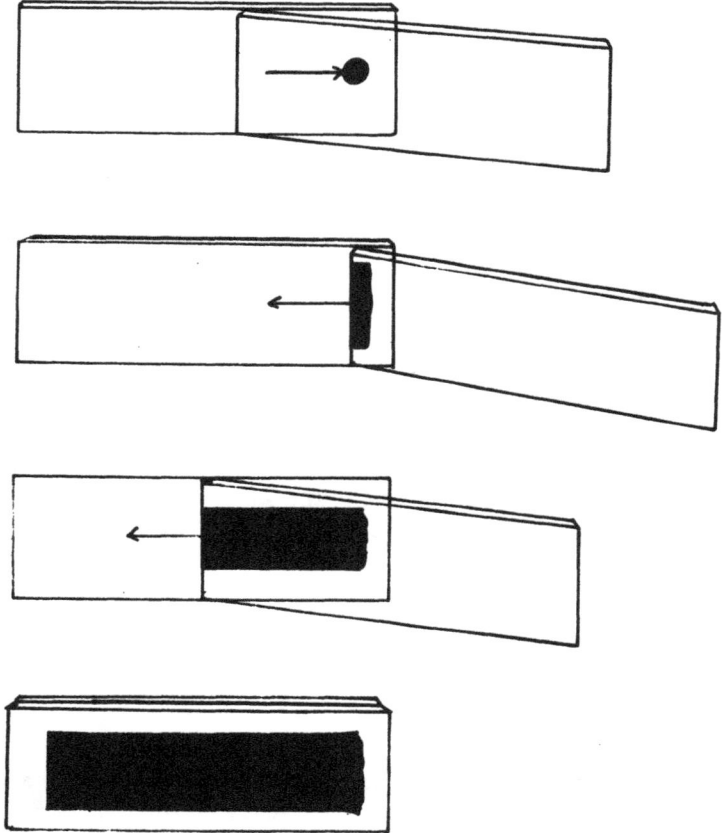

Fig. 52.   Technique of preparing a blood smear with two glass slides (described in the text).

Pull back the upper slide until its edge touches the drop of blood. Capillary attraction will then cause the blood to spread along the edge of the inclined top slide.

Hold the lower specimen slide firmly with one hand and push the upper slide gently and steadily forward until it reaches the opposite end of the lower slide.   Finally remove the upper slide.

Do not exert unnecessary pressure when pushing the upper slide forward, since this may damage the blood cells. It should be observed that the greater the angle of the upper slide, and the slower the forward movement of the upper slide, the thicker the film.

Allow the slide to dry at room temperature in a clean atmosphere. If desired, the smear may be dried and fixed by passing the bottom of the slide two or three times over a flame. The temperature of the slide should not exceed 80°C.   A good film will be even and smooth.

If the slides are to be stained, the staining should be done within 24 hours.

STAINING BLOOD SMEARS OR FILMS — WRIGHT METHOD

Stained blood smears are advantageous in that certain parts of the blood cells are more clearly differentiated for better study. Staining is also required for the diagnosis of certain blood diseases.   Wright's stain is used routinely for staining blood smears (see Appendix).

The general procedure for staining blood films is as follows :

Step 1.   Prepare a thin, fresh blood smear as described above, and dry thoroughly by waving the slide in air.

Step 2.   Place the slide on a flat surface.   If the smear is on a cover glass, use a cork to hold the cover glass during staining.

Step 3.   Apply Wright's stain to the dried film and allow to remain for 1 minute. The volume should be just sufficient to cover the specimen completely. (Another method is to place the slide over the mouth of an open Coplin jar and flood it with Wright's stain. Special staining racks are also available.) ·

Step 4.   Add the phosphate buffer solution (see Appendix) drop by drop over the stain.   The quantity should be sufficient to cover the stained specimen without running off the slide. Allow to remain on the slide for 2 to 5 minutes; the actual time can be determined by trial and error.

Step 5. Wash the excess stain from the slide completely with a jet of water from a wash bottle. After washing for 5 to 10 seconds, the blood film as a whole should appear lavender pink. The water acts as a destaining agent and differentiates the colors within the specimen (see page 138 for a description of differentiation in staining). Excessive washing of the slides will completely remove the stain.

Step 6. Dry the slide at first by tilting slightly and applying a piece of filter paper to its edge, then by waving in air or by placing the slide between blotting papers and pressing very gently without pushing the slide in any direction. Repeat the blotting procedure with dry sections of the blotters until the slide is thoroughly dry.

Step 7. The slide may be either mounted or unmounted for examination. If it is to be mounted, mount in neutral balsam or, preferably, in a synthetic mountant such as clarite.

Step 8. Examine the slide using the oil-immersion objective. If the slide is not mounted, apply the immersion oil directly to the surface of the dried film.

Seen microscopically, the cells of a good stained slide should show well-defined stained cells without blurred edges and with clear areas around the cells. The red corpuscles will be buff color, not yellow or red. The nuclei of the various white cells should show mainly as purplish-blue. The platelets will be stained purple-blue and will be sharply defined, while the eosinophils will appear bright red.

If the slide is to be stored, remove the immersion oil from it with a little xylene or benzene.

BACTERIOLOGICAL SMEARS

New slides or thoroughly cleaned slides must be used for the preparation of bacteriological smears, since otherwise false results may be obtained.

Step 1. Sterilize an inoculating needle or loop (see Fig. 53) in a

Fig. 53. Inoculating needle (top) *and loop* (bottom).

flame and touch it to the surface of the liquid culture. Transfer the needle or loop to the center of the glass slide and smear the culture lightly over an area of about 1 cm². Do not transfer too much of the culture, since observation and identification of the bacteria under the oil-immersion objective will thereby be impaired.

If a smear is to be made of bacteria that have been cultured on a solid medium such as agar, it will be necessary to dilute the specimen on the slide. The procedure is to place a drop of sterile water on the center of the slide and then to add a loopful of the culture to the water. The emulsion is mixed thoroughly and spread over about 1 cm².

Step 2. Allow the smear to dry at room temperature.

Step 3. If the dried smear requires fixing, hold the smear downwards and pass the slide rapidly once or twice through the flame of a Bunsen burner. Finally, flame the underside of the slide. Do not use too strong a flame. Fixing may also be accomplished by immersing in methyl alcohol or other chemicals. The bacteriological smear is now ready for staining.

STAINING BACTERIOLOGICAL SMEARS

Staining of bacteriological smears is required in the detection, examination, and identification of bacteria. Smears must be stained to bring out cell structure and also in the study of bacterial reaction to special stains. An example of a stained bacterial preparation is shown in Fig. 54.

The general procedure for staining a bacterial smear or film is as follows :

Step 1. Prepare the smear as described above, using the smallest quantity necessary for the purpose.

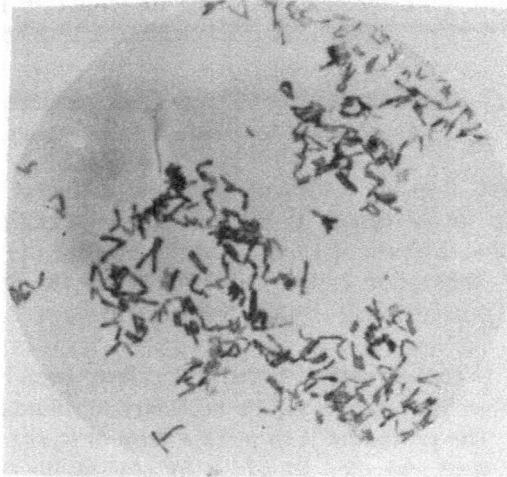

Fig. 54. Photomicrograph of a stained preparation of diphtheria bacilli.

Step 2.   Place a drop of the stain (crystal violet, for example) on the flamed smear and allow to remain from 30 to 60 seconds.

Step 3.   Wash off the excess stain with a jet of water from a wash bottle.   Allow the slide to dry.

Step 4.   Mount in a neutral balsam preparation if the slides are to be preserved.

Step 5.   Apply a drop of immersion oil to the mounted or unmounted smear and examine with the oil-immersion objective.

Gram's Method

Gram's method is valuable for the differentiation and examination of different species of bacteria.   A given bacterial microorganism may be Gram-positive, Gram-negative, or Gram-variable.   It often happens that many Gram-positive bacterial cells become Gram-negative because of unusual environmental or physical conditions, such as temperature of incubation, degree of acidity of the medium used, age of the culture or breakdown of the cells in their own serum (autolysis).

Apparently the protoplasm of the bacterial cell contains magnesium ribonucleate ; the presence of this compound is associated with the Gram-staining property of the cell.

Bacteria, when treated with Gram's iodine solution, act as a mordant. Gram-positive organisms will retain a crystal violet dye even though decolorizing agents, such as ethyl alcohol (95%), may be added later. Gram-negative organisms do not retain the violet stain when a decolorizing agent is added, but can be subsequently stained with another dye (safranin, basic fuchsin or methylene blue).

The general procedure in Gram's method is as follows :

Step 1. Prepare the smear as described above using the smallest quantity necessary for the purpose.

Step 2. Place a drop of the crystal violet solution on the smear and allow to stand for 2 minutes (see Appendix for the method of preparing the Gram stain).

Step 3. Wash off the excess stain by rinsing with a large volume of water.

Step 4. Apply a few drops of the iodine solution (see Appendix) to the slide and allow to stand for 1 minute or place the slide in a Coplin jar containing the iodine solution.

Step 5. Rinse the slide in water to remove the excess iodine solution.

Step 6. Decolorize in alcohol (95%) for about 30 seconds or until the alcohol flowing from the slide is not colored with the violet dye. Gently agitate the alcohol over the slide during the alcohol application. As soon as the alcohol is free from color, wash well with water to stop the staining process. An alternate decolorizer is acetone or a mixture of equal parts of alcohol (95%) and acetone.

Step 7. Apply a counterstain, using a solution of safranin (see Appendix), and allow to remain for 10 seconds. The counterstain provides a contrasting color, so that the gram-negative organisms can be clearly seen.

Step 8. Wash off the excess counterstain with water. Dry the slide without the use of blotting material.

Step 9. Mount the specimen in balsam (see page 140 for the mounting procedure).

Examine the slide under the microscope using the oil-immersion objective. Gram-positive organisms stain blue, Gram-negative, red.

Spore Stain

This type of staining differentiates between bacterial cell bodies and spores. The staining procedure is similar to that for Gram's method, with a few modifications.

Step 1. Prepare and fix a bacterial smear as described above.

Step 2. Flush the slide with a carbol-fuchsin stain (see Appendix).

Step 3. Wash the slide with hot water and rinse with alcohol (95%).

Step 4. Counterstain with Loeffler's methylene blue stain (see Appendix) and allow to act for 2 to 5 minutes.

Step 5. Wash the slide in water and allow to dry. Examine under the microscope using an oil-immersion objective.

The spores are stained red and the cell bodies are stained blue. When Loeffler's methylene blue is used alone without a counterstain, the metachromatic granules of the cell stain intense purple and are differentiated from the cytoplasm, which stains blue. This procedure is used to examine throat cultures for diphtheria bacilli.

Acid-Fast Stain—Ziehl-Neelson Method

The acid-fast stain is a general bacteriological stain that differentiates between acid-fast bacteria and those that are not. It is used to detect and differentiate the acid-fast bacteria of tuberculosis, leprosy, and other diseases.

Step 1. Prepare a fixed smear as described above. The sample for demonstrating the tubercle bacilli may be collected from a sputum specimen. Yellow particles seen in the sputum may be assumed to harbor the tubercle bacilli and may be taken as the sample.

Step 2. Flood the specimen with a prepared carbol-fuchsin stain (see Appendix). Hold the slide over a low flame and steam gently, but do not boil, for about 3 minutes. If evaporation occurs, replenish the stain on the slide. The heating is necessary, since under normal conditions, it is difficult to stain the protoplasm of acid-fast bacilli.

Step 3. Wash the slide in running water.

Step 4. Decolorize the slide with an acid-alcohol solution (made of 1% hydrochloric acid in 70% alcohol) until the alcohol does not retain or extract the dye. Wash the slide again with water.

Step 5. Counterstain the specimen with Loeffler's methylene blue for 1 minute.

Step 6. Wash in water, dry thoroughly, and mount in balsam.

Step 7. Examine the slide under a microscope using the oil-immersion objective. The acid-fast bacteria are stained red, the other bacteria blue.

Capsule Stain—Hiss Method

Certain bacteria are surrounded by capsules useful in the identification of unknown species. Such capsules can be seen more clearly against a stained background.

Step 1. Prepare a bacterial smear and allow to dry in air.

Step 2. Apply an aqueous solution of crystal violet (1%) to the smear and steam gently over a flame for half a minute.

Step 3. Wash off the stain immediately with an aqueous solution to copper surfate (20%).

Step 4. Allow to dry and mount in balsam.

Step 5. Examine under a microscope using the oil-immersion objective. The capsules are composed of polysaccharides, which do not stain easily and are thus seen as holes against a dark background.

PREPARING A DRY MOUNT—CELL MOUNT

Cell mounts are generally prepared by confining the specimen within a circular area on the slide. They are used for fluid, semifluid, and balsam mounts.

The cell is prepared by affixing to the slide a solid cementing ring large enough to confine the specimen. Materials used include metal and cardboard washers and plastic rings. Rings made of liquids that harden when dry, such as shellac, are also used.

The washers or rings are cemented to the exact center of the glass slide with household cement or Canada balsam. (To find the exact center of the slide, draw the diagonals on a 75-by-25-mm card; since the diagonals intersect exactly in the center, the card can be placed under the slide to show the center.) A typical cell mount is shown in Fig. 55.

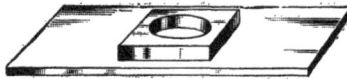

Fig. 55.   Cell mount.

When the washer adheres rigidly to the slide, place the thoroughly dried specimen in the hole of the washer and cement a cover glass to the top of the washer. The specimen must be thoroughly dry; otherwise, cloudiness may appear below the cover glass or fungus growths may form.

The thoroughly dried cover glass is ringed with cement before it is placed over the cell. It is then carefully positioned and allowed to set firmly. A final ring of cement is placed around the edge of the cover glass, simultaneously sealing off the specimen and providing a final finish for the slide.

If a deeper well is desired for a thicker specimen, two or more washers are cemented together. If doubtful whether the cover glass is thoroughly dry, take the cover glass with forceps and pass it several times through a medium flame.

To immobilize a specimen in the cell, place a minute amount of gum tragacanth in the center of the cell and allow to dry; breathe on the gum to make it slightly tacky and place the specimen in the gum, positioning it with a needle.

Another method of preparing shallow mounts is to paint a circle of shellac, black varnish, or gold size directly on the slide. (To paint a circle freehand, fasten a metal washer to the glass slide with balsam; reverse the slide and paint the circle with a fine-pointed brush, using the washer as a guide.) Allow the paint to dry. If a deeper well is desired, successive layers of the shellac can be applied. The washer can be removed with a little xylol.

A turntable (see Fig. 56) can be used to prepare more profess-

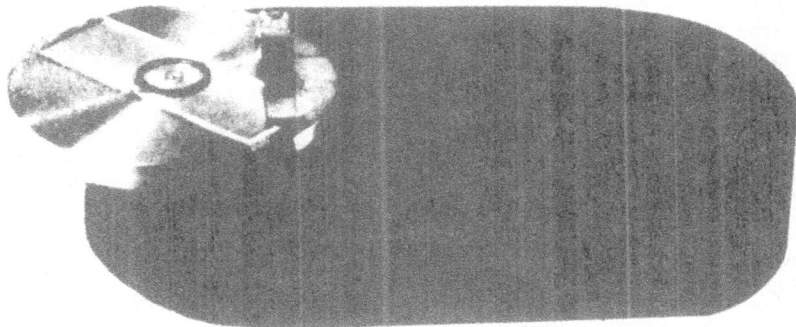

Fig. 56.   Turntable for preparing cell mounts.

ional-looking cell mounts. Such a turntable, generally of metal, can be bought or made in that laboratory. It consists of a circular disc or solid wheel, about $3\frac{1}{2}$ inches in diameter, which can be rotated with the fingers. Attached to the turntable is a hand rest to support the hand while holding the brush against the slide, which is clipped into position. A centering gauge serves as a guide. The diameter of the circle should be slightly less than the width of the cover glass.

In using the turntable, position the glass slide by means of the clips. Dip a fine-pointed brush into the shellac, varnish, or gold size. Place the hand holding the brush on the hand rest. With the other hand, start the wheel rotating. Lower the brush carefully so that the tip of the brush touches the desired radius on the slide. Allow the brush to remain on the slide for several revolutions until a layer of fluid is built up. Dry the slide in an oven; then apply another layer over the original ring. Repeat until the desired thickness is attained. Warm the slide to drive off any moisture occluded in the shellac or other substance.

Mount the specimen in the cell and then place the slide again on the turntable. Press a thoroughly dry cover glass in position. Rotate the turntable and apply several coats of shellac around the edge of the cover glass to seal it securely to the slide. Allow the slide to dry thoroughly.

MOUNTING DIATOMS

The method of mounting diatoms depends on whether the shell structure only, or the complete organism with its chromatophores, is to be mounted.

Mounting Diatom Shells

Place a few drops of the culture on a cover glass and allow the water to evaporate at room temperature. Place the cover glass in a wide porcelain crucible. With tongs, hold the crucible steady in a low Bunsen flame. The heat will destroy the content of the cell and any remaining organic matter. Under the flame, the diatoms will turn brown, then gray-white. Do not keep the cover glass in the flame too long since the soft glass will start to melt if it is subjected to intense heat. Allow the glass to cool and examine under low-power to determine whether all organic matter has been destroyed.

Make a suspension of the diatomaceous ash by transferring it to a glass vessel, adding water, and shaking. Transfer a few drops of the suspended material to a clean cover glass resting on a flat surface and let the drops spread into a thin film. Allow to dry in a dust-free atmosphere or hasten drying with heat.

In the center of a clean glass slide, place a drop of balsam or, better still, a resin of high refractive index, which will give better resolution. Invert the cover glass over the slide so that the diatoms fall into the balsam. Allow the slide to dry, then prepare a ring around the specimen as described above. Place a coverslip over the ring mount and seal with balsam or cement.

Slides can be prepared with many varieties of diatoms arranged in beautiful geometric patterns. Mounting diatoms in this way has become a hobby of many microscopists. When viewing a slide prepared in such a manner, one is amazed at the fantastic geometric configurations that can be achieved in mounting individual minute microscopic organisms.

It is necessary, at first, to separate and isolate the different diatom shells by placing a drop of the prepared shell sediment on a slide. While viewing the different specimens under the microscope, the operator picks up the individual shells and drops each kind in a separate labeled jar containing distilled water. A very fine-pointed brush can be used, but more often a single hair from a shaving brush or a single cat's hair fastened to a handle with adhesive or cellophane tape is used for handling diatoms.

The individual diatom may be directly mounted on a slide or cover glass in any geometric pattern desired. The area to which the diatoms are to be applied is coated with gum tragacanth. Before transferring the specimens to the slide, moisten the gum by breathing on it. This procedure will fasten the diatoms to the slide. The pattern in which you wish to arrange the diatom shells within the ring is temporarily drawn in ink on the underside of the slide or cover glass. Another method is to sketch the geometric pattern on a piece of paper, which serves as a guide as the worker looks through the microscope and mounts each shell on the slide or cover glass. Diatomaceous earth, sold commercially as an abrasive, is a good source for diatom shell specimens.

Mounting Complete Diatoms

When mounting complete diatoms so that the cell structures can be studied, they must be fixed prior to mounting.

Mix the cleaned diatom sediment with the fixing solution in a beaker. Bichromate-acetic acid fixer (see Appendix) is added in the ratio of 50 parts by volume of fixer to 1 part of the diatom sediment suspension. Allow to stand for several hours or overnight. Decant the fixing solution from the sediment, replace with water, and stir frequently. Allow to settle again and pour off the water. Add fresh water and stir again. Repeat the washing procedure four times. The diatoms are then placed on a slide, stained, dehydrated, and mounted in the described manner (see page 140).

A substantial quantity of diatoms must be on hand before mounting is begun, since many are lost during mounting. A centrifuge can be used for concentrating the diatom culture.

PREPARING A PERMANENT BALSAM MOUNT

To protect and preserve a specimen, it must be sealed in a liquid or solid medium called a *mountant*. The various steps necessary to prepare a wax section of a specimen that is to be permanently mounted on a slide will therefore be described. A few of these steps can be omitted if the specimen is mounted whole, without recourse to section cutting. Before mounting any live organisms, such as algae, protozoa, higher invertebrates, or plants, a similar procedure must be followed to prevent distortion of the specimen.

Sections are then sliced from a prepared large organic specimen, cut if possible to a single cell thickness, so as to allow microscopical examination by transmitted light.

Fixation and Hardening of the Specimen

Fixation of the specimen is an important first step in preparing adequate microscopical specimens. The prime purpose of fixing the specimen is to preserve the specimen and its parts, such as the cells and nuclear elements, so they will approximate the form of the killed animal or plant.

Most fixing solutions stop decay and decomposition of the specimen and simultaneously initiate the hardening of the cells that continues during the dehydration process to be described. The hardening process strengthens the cell walls of the specimen and gives the tissues the necessary texture for cutting good sections.

Fixation of the specimen must be done as quickly as possible on the dead animal tissue or plant, and is generally carried out at room temperature. In a few cases where haste is necessary, the specimen can be boiled in the fixing solution.

The length of time required to fix a specimen depends on the thickness of the specimen and the type of fixing solution used. Generally speaking, thinner specimens allow the fixer to penetrate more rapidly. From 24 to 48 hours are sufficient for adequate fixing of the average specimen.

The usual procedure is to place the specimen in a 4-oz screw-capped jar (or smaller, if the size of the specimen permits). Cover the specimen with a generous amount of the fixing solution. (The chemicals may be corrosive or may give off irritating odors, so wear rubber gloves, especially when working with formol-saline solution.)

A 10% formol-saline solution (see Appendix) is a general fixative for routine work. Strong alcohol solutions, at least 70%, are also used for fixing specimens. Other fixatives are available for special tissue work.

After fixing in a formol-saline solution, the specimen must be thoroughly washed for a half hour or an hour in running water to remove all remaining chemicals.

Dehydration of the Specimen

For many reasons, it is imperative to remove the water from a fixed organic specimen. Organic specimens, both plant and animal, contain water in their cells. When the specimens are exposed to air, the water evaporates through the cell walls, causing the specimens to dry out and the cell walls to collapse; therefore, the water must be removed if the specimen is to keep its original shape.

Dehydration of the specimen is essential when preparing wax or celloidin sections. Water prevents the tissue from being totally impregnated with the wax or celloidin, and most of the media in which the specimens are to be mounted are immiscible in water.

Basically, the dehydration process consists of treating the specimen with a series of solutions containing the dehydrating agent in progressively increasing concentrations, a treatment that progressively decreases the concentration of water in the specimen. The dehydrating agent most generally used is either ethyl, methyl, or isopropyl alcohol. Ethyl and isopropyl alcohols are much safer (the vapor of methyl alcohol is toxic). The alcohol not only removes the water but also prevents decomposition and strengthens the cell walls.

The reason for carrying out dehydration in stages, using increasing concentrations of alcohol in water, is that if the specimen were dipped in absolute alcohol, the cell walls would become unduly hardened, entrapping some of the water within the cell and eventually causing decay. If, however, a dilute concentration of alcohol (20%) is employed, the alcohol will replace some of the water without hardening the cell wall. As the concentration of the alcohol is increased, the water is gradually replaced with alcohol, and thus all the water is removed before total hardening of the cell wall occurs. In this way, the cell is not only dehydrated, but is also fixed and hardened.

The general method of dehydrating the specimen, which is done not only after fixing but also after staining, is as follows :

Step 1. Put the specimen into a 20% alcohol solution for 1 to 6 hours, the time depending on the size and penetrability of the specimen.

Step 2. Replace the 20% alcohol solution with a 30% solution and keep the specimen in this solution for 1 to 6 hours (use forceps to transfer the specimen). Another method is to leave the specimen in the old solution, decanting this solution from the specimen and then promptly replacing it with the solution of the higher concentration. The first procedure is preferred, however, since complete dehydration can be better achieved. Care must be taken that the specimen is at no time allowed to become dry between changes of the solutions.

Step 3. Replace the 30% alcohol solution with a 50% solution and keep the specimen in this solution for $\frac{1}{2}$ to 6 hours. Do not keep the specimen in the solution longer than necessary, since in some cases, the long dehydration period may cause softening of the tissues, excessive shrinkage, and distortion of the cells.

Step 4. Replace the 50% alcohol solution with a 75% solution and keep the specimen in the solution for ½ to 6 hours.

Step 5. Finally, replace the 75% alcohol solution with absolute alcohol or a 95% alcohol solution and allow the specimen to remain from 1 to 12 hours. This step may be optional for many tissue specimens if the specimen can be later cleared of the 70% alcohol with xylene or aniline.

Tertiary butyl alcohol dissolved in various concentrations of ethyl alcohol is used extensively as a dehydrating agent. It has been found to be excellent, since the reagent has a lesser tendency to cause the tissues to become hard and brittle. This reagent has the disadvantage, however, of solidifying at 25°C and, with its low boiling point (83°C), may be a fire hazard. *p*-Dioxane is another good dehydrating agent, since it is soluble in water, paraffin solvents, and mountants. *p*-Dioxane is quite miscible with water and therefore easily replaces the water in the tissues. The substitution in this case is not accompanied by unusual stresses, and thus faster dehydration occurs, requiring fewer operations and therefore producing less distortion of the specimen. Dioxane vapors are toxic, however, and have proved to act as a cumulative poison. Dioxane should be used in a well-ventilated room or under a hood. The three progressive solutions required for complete dehydration are: (1) 30% *p*-dioxane in water, (2) 60% *p*-dioxane in water, (3) anhydrous *p*-dioxane.

Dealcoholization and Clearing of the Specimen

After dehydration, the alcohol must be removed from the specimen, since the alcohol is incompatible with both paraffin wax (wax section) and balsams (in the case of permanent mounting). In clearing the specimen, the alcohol is replaced with an organic solvent that is freely miscible with the alcohol and, at the same time, acts as a solvent for the wax or mountant. The clearing agent, moreover, enhances the transparency of the specimen.

The chemicals most often used for clearing the specimen are toluene, xylene, and benzene. Xylene (xylol) is most often used in routine microscopy; it is cheap and fast in action, it makes most tissue sections transluscent or transparent, and it has the additional property of hardening tissues. It is miscible with absolute alcohol, balsam, and melted paraffin wax. Xylene does, however, have a few disadvantages. It is immiscible with water, and, indeed, turns milky in the presence of water; this property may be advantageous, however, in that it reveals incomplete dehydration of the sample. Since xylene may also cause brittleness in some tissues, other clearing agents, such as chloroform, aniline, or cedar-wood oil, are used. Benzene is similar to xylene in its action as a clearing agent, although it is less hardening. It is, however, inflammable and highly toxic.

The length of time a specimen should remain in a clearing fluid depends primarily on its thickness and whether the clearing agent can permeate the tissues easily. When xylene is used, the time may vary from 15 minutes to 2 hours. When a tissue is completely cleared, it will appear transluscent. Do not keep the specimen in the xylene longer than necessary, especially when the specimen is to be infiltrated with paraffin wax. Many workers use a progressive series of concentrations of xylene in alcohol (25, 50, 75, and 100%).

Infiltration of the Specimen with Molten Paraffin Wax

A specimen prepared for section cutting must be reinforced and supported so that when slices are cut, they will not be crushed or torn in the process. The general method of reinforcing specimens is to impregnate the tissues completely with molten paraffin wax. The hot, liquid wax replaces and slowly eliminates the clearing reagents in the specimen and then becomes solid when it cools.

Paraffin waxes for this purpose must be smooth, of even texture, and free from dirt and occluded water. Commercial paraffin waxes with differing melting points are available. For routine work, a wax that melts at 58°C is adequate.

The wax is placed in a metal or borosilicate glass receptacle and heated gently until the wax melts. The wax is kept liquid in a thermostat oven, if available, kept at about 3 or 4°C above the melting point of the wax.

The procedure for infiltrating the specimen is as follows :

Transfer the specimen to the molten paraffin bath for 1 to 4 hours, depending on the time necessary to eliminate the clearing from the tissues. The thicker the specimen, the longer the interval of time that will be necessary for its infiltration. A good practice to hasten infiltration and remove the clearing agent is to replace the initial molten wax saturated with the clearing agent with fresh molten wax. This is done by transferring the specimen, using forceps, to a fresh dish containing molten wax. Impregnation with molten wax can also be done in a vacuum oven, a procedure that reduces the time by one-half. After infiltration, the wax is cooled and the specimen is allow-ed to remain in the wax until ready to be embedded.

Embedding the Specimen

After the specimen has been thoroughly infiltrated with the liquid wax, the next step is to embed it with fresh paraffin wax having the same melting point. This procedure is also called *blocking*, since the specimen and the molten wax are isolated by casting into a mold.

There are various types of embedding molds. These molds can be assembled and disassembled, and can be increased or decreased in volume by interlocking devices. The type of mold generally used, which gives excellent results, is made of two L-shaped bars of metal, generally brass or stainless steel, about 2 by 1 by 3/4 in. (see Fig. 57). These L-bars, which can be made or purchased,

*Fig. 57.   L-bar embed-
ding mold.*

are readily adjustable to specimens of various sizes. They are placed on a brass, stainless-steel, or porcelain plate when the specimen is ready to be embedded.

MAKING A PAPER BOAT

If L-bars are not available, molds can also be made from stiff, strong paper. To make a paper boat, fold a rectangular piece of heavy glazed paper or aluminum foil in thirds; then fold each end of the paper to give a depth greater than the width (see Fig. 58, *top*).

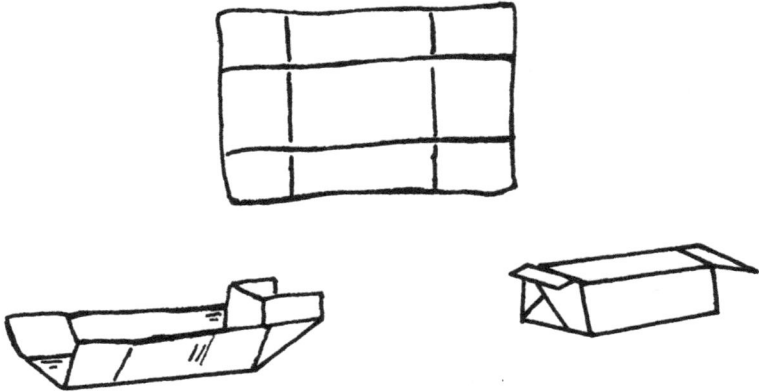

Fig. 58. Method of folding a paper boat (described in the text).

Each end of the boat, when folded upright, will consist of a middle part and two tab ends (Fig. 58, *left*). Turn the tab ends toward each other and then turn down or crimp the upper part over these tab ends so that they are held in place; repeat this procedure with the other end (Fig. 58, *right*). The boat should be sufficiently large to hold the specimen loosely. With either the L-bar mold or the paper boat, a thermostatically controlled hot plate or oven is used to keep the paraffin molten until the specimen is embedded.

The general procedure for embedding the specimen in wax is as follows :

Step 1. Melt the paraffin wax and filter it through cotton into a borosilicate glass beaker. Place the wax on a hot plate or in an oven to keep it molten until ready for use.

Step 2. Adjust the L-bar mold on a porcelain or metal base so that the area will be slightly larger than the specimen. Fill the mold with the molten wax.

Step 3. Gently warm the container holding the infiltrated specimen and the wax over a Bunsen flame until the wax is melted, applying only enough heat to liquefy the wax.

Step 4. Warm the tips of a pair of forceps and use them to remove the specimen quickly, so that the wax will not solidify. Place the specimen, with the surface to be cut facing downward, in the mold containing the molten wax.

Step 5. Push the specimen gently into the wax, using a warm needle. Fill the mold with additional molten wax. If necessary, the specimen can be oriented and centered with warmed needles.

Allow the wax to harden at room temperature. If the type of paraffin wax used has a tendency to crystallize at room temperature, it should be cooled quickly as soon as it is sufficiently hardened, by plunging it into a receptacle containing cold tap water or ice water.

Step 6. Examine the congealed wax specimen. If large, white, opaque spots are present near the specimen, they are doubtless due to xylene or alcohol still present and indicate that infiltration of the specimen is not complete. Repeat the infiltration and embedding process. The top surface of the mold may appear concave (indicating shrinkage), but this defect may be disregarded unless the specimen is insufficiently covered with wax.

Cutting Paraffin Wax Sections

Cutting wax sections requires a skill that can be acquired with a little practice. There are various tools or instruments used in the cutting process. A few of the more common ones are described below.

MICROTOMES AND ACCESSORIES

TABLE MICROTOME

The table microtome (see Fig. 59) is a reasonably priced instrument, adequate where speed is not essential and where a section not less than 5 micrometers in thickness is satisfactory. The specimen is placed in a clamp holder, which is activated by a feed screw device that forces

Fig. 59.   Table microtome.

the specimen upward.   The screw device is controlled at its lower end by a knob graduated in 5-micrometer intervals.   One complete revolution of the knob cuts a section 0.5 mm thick.   Two horizontal glass plates on top of the instruments provide a surface on which a knife slides smoothly for uniform cutting of the specimen.   The main frame is provided with a heavy clamp at the back for securing the instrument to the table.

MICROTOME KNIFE

A sharp knife with a handle (see Fig. 60) is a necessary accessory of the microtome.   (A straight-edge razor can also be used for cutting sections.)

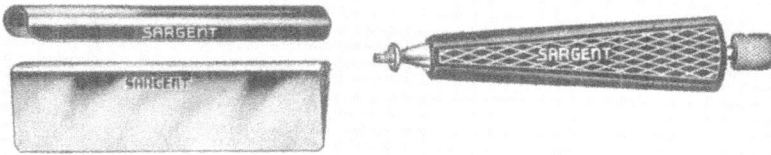

Fig. 60.   Microtome knife (blade, back, and handle).

ROTARY MICROTOME

The advantage of the rotary microtome (see Fig. 61) is that the

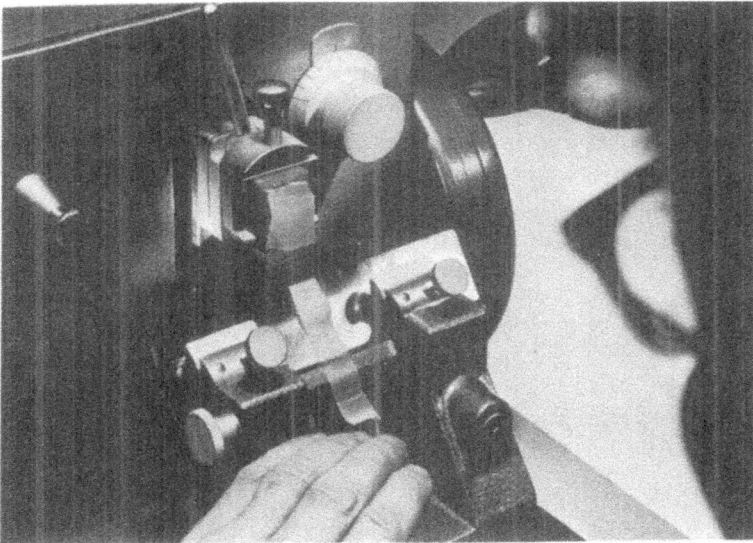

Fig. 61.   Rotary microtome.

specimen is fed automatically to the knife by means of a hand crank. The knife is held in a stationary position and the tissue specimen is moved up and down past the cutting edge, which operates the same way as a wood plane.   The feed adjustment, which controls the cutting thickness, can be set for any thickness, generally from 1 to 50 micrometers, in 1-micrometer steps.   A knife holder is provided with a clamp to hold the knife rigidly.   It can be tilted through 30° of cutting angle.

The knife is an integral part of the microtome, and a microtome is only as good as the cutting edge of its knife. Microtome knives are made of tempered steel, ground, polished, and honed to give a sharp, durable cutting edge. They are available in several sizes (110, 120, 185, and 280 mm) and are usually sold with a fitted back and handle to facilitate honing and stropping.

SHARPENING THE MICROTOME KNIFE

It is necessary to hone and strop the knife from time to time to keep it razor-sharp, as well as to remove any nicks in the cutting edge of the knife. A knife with a poor edge will not cut good ribbon sections and will produce poor mechanical compressed sections. If you are uncertain as to whether the knife is sharp and its cutting edge is free from nicks, examine it under the microscope by reflected light, using 100X magnification. Under the microscope, a good cutting edge should appear as a narrow, straight line, free of scratches and practically free of nicks and serrations. If only a small part of the knife edge is dull, it would be advantageous to use the remainder of the edge before resharpening.

Honing

Many kinds of microtome knife hones, or oil stones, are sold, and they can be bought in various grades of fineness. The yellow Belgian knife hone is considered to be a good quality oil stone. It has exceptionally fine grit and uniform texture to give fine cutting properties. Another type of sharpening stone in common use is the carborundum stone.

If a stone is not available, a piece of beveled plate glass about 14 in. long, 1/4 in. thick, and with a width slightly greater than the length of the blade, is a good substitute. The abrasive used with the glass plate consists of a paste made with levigated powdered aluminum oxide and water.

Stone hones usually come in wooden trays with covers (see Fig. 62). When not in use, they should be kept covered. Before use the hone must be lightly covered with a lubricant—the commonest are thin petroleum oils, olive oils, xylene and soapy water.

To hone the blade, proceed as follows:

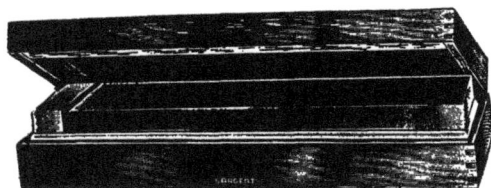

Fig. 62.   Stone hone.

Step. 1   Remove any dust or dirt from the surface of the hone with a cloth moistened in benzene or xylene; let it dry.

Step 2.   Spread a thin film of oil over the stone. (If soap and water are to be used as lubricant on a carborundum stone, soak it in water for about 5 minutes before applying the soapy film.)

Step 3.   Attach the handle and fitted back to the knife. Hold the ends of the knife in both hands and lay the knife flat on one end of the stone in an oblique position. Keep the blade perfectly flat by using a little uniform pressure with the fingers of both hands on the knife proper. Using a steady, slanting stroke, with sufficient pressure to hold the knife uniformly against the hone, push the knife forward, *edge first,* to the opposite end of the hone (see Fig. 63, *left*). (*Note that the knife must be kept perfectly flat*; *if it is raised even slightly, the edge of the knife will be damaged.*)

When the blade reaches the far end of the hone, turn the blade over by using the back of the knife as a pivot and move it again across the hone, still using a slanting stroke (Fig. 63, *right*). Usually, 5 to 10 strokes will be sufficient for new knives. The technique of honing on a glass plate is similar to that of honing on a stone, except that the knife, held flat as on the stone, can be pushed straight, rather than obliquely (see Fig. 64). Also, of course, abrasive paste, rather than a lubricant, is used.

The progress of the honing can be judged from the reflection of the light coming from the knife edge. When these reflections have almost disappeared, examine the knife under the microscope (100X)

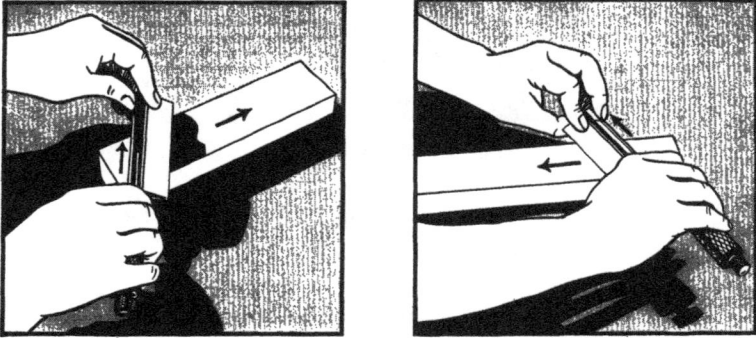

Fig. 63. (above) *Technique of honing the microtome knife on a stone (described in the text).*

Fig. 64.   *Technique of honing the microtome knife on a glass plate (described in the text).*

to see whether the nicks have disappeared and the edge of the knife appears smooth.

Step 4. Wipe the knife with a cloth moistened with benzene or xylene and let it dry; wipe the hone with xylene, let it dry, and store it in its original case. If the cutting edge of the knife has large nicks that cannot be eliminated by honing, it should be reground by a specialist in grinding microtome knives.

Automatic Microtome Knife Sharpener

The automatic microtome knife sharpener (see Fig. 65) locks the knife in place at a fixed angle. The edge is then stroked automatically against a vibrating glass honing plate smeared with abrasive. After several strokes, the knife is automatically turned over and the other

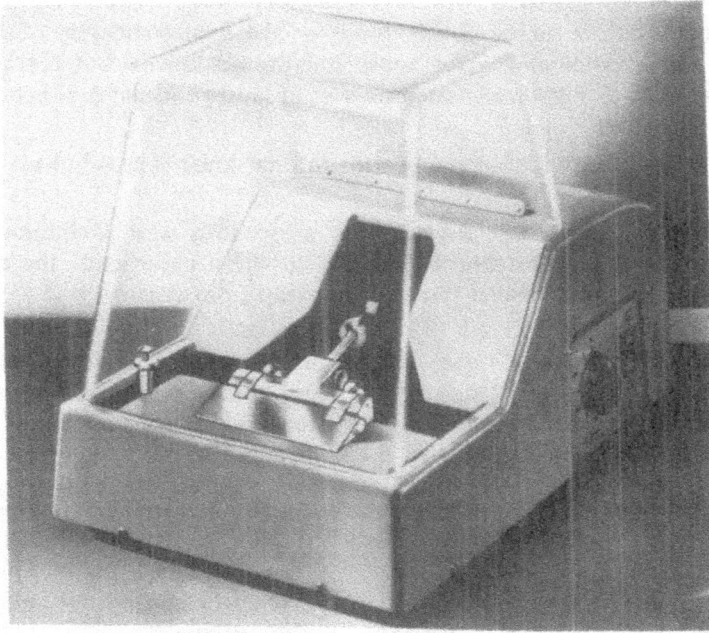

Fig. 65. Automatic microtome knife sharpener.

side is honed by a number of strokes.   The duration of the stroking cycle is set by an electrical timer.

Stropping

The final operation in sharpening a microtome knife is stropping. The commonest type is similar to those found in barber shops, although they are especially designed for microtome knives.   The strop consists of two kinds of stropping leathers bonded together, 24 by $2\frac{1}{2}$ in.   There is a handle at one end and a device on the other for attaching the strop to a hook (see Fig. 66).

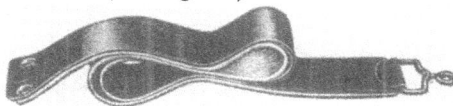

Fig. 66.   Microtome knife strop.

One of the leathers, on the side used for initial stropping of the knife, is impregnated with fine diamond dust. The other is fine pigskin with a matte finish, used for the final stropping. Such a strop is recommended for most microtome knives, but for larger knives a leather strop mounted on a solid hardwood block is better.

The procedure for stropping a microtome knife is as follows :

Step 1. Attach one end of the strop to a wall or immovable object at a comfortable height. Hold the other end of the strop rigidly in the left hand to prevent the strop from sagging.

Step 2. With the right hand firmly grasping the knife handle, place the blade flat at one end of the strop in an oblique position. Draw the knife forward along the strop with the movement of the stroke away from the edge of the knife (see Fig. 67, *left*). *Note* that *this is the reverse of the honing movement.* At the end of the stroke,

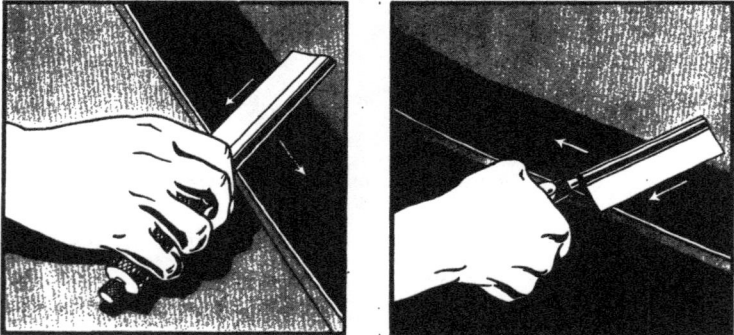

Fig. 67. Technique of stropping the microtome knife (described in the text).

turn the knife over and repeat the movement (Fig. 67, *right*). Do not move the knife rapidly, but smoothly and in rhythm. As in honing, do not lift the edge of the knife, since this will round the edge. From 15 to 20 strokes should be sufficient.

Step 3. Turn the strop over and strop the edge on the fine pigskin. From 15 to 20 strokes should suffice. Wipe the knife clean with a soft cloth.

PREPARING THE WAX BLOCK FOR CUTTING

The procedure for preparing the wax block for cutting by a slide or rotary microtome is as follows :

Step 1.   Remove the paraffin block from the mold and trim it until the thickness of the wax is about 1/4 in.

Step 2.   Using a knife or razor, trim more wax from the two sides of the wax block that will be parallel to the microtome knife until they are about 1/8 in. from the specimen proper.   Pare the wax from the other two sides until the specimen is almost laid bare.

Step 3.   Coat the specimen holder with wax.   Plastic, wood, and metal specimen holders are available for paraffin-embedded specimens that are to be mounted in the clamp attached to the microtome. Metal mounting discs (see Fig. 68) are preferred, since they are more rigid and allow very thin sections to be cut.   The object holders used for rotary microtomes are generally round and grooved, so that the wax block adheres more firmly to the grooved surface.

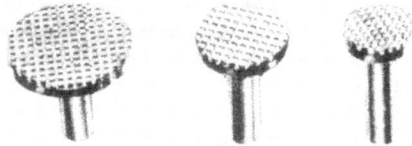

Fig. 68.   Rotary microtome specimen holders.

Step 4.   Heat a metal knife or spatula in a Bunsen flame until it is hot enough to melt the paraffin wax.   Immediately place the flat side on the surface of the specimen holder.   Simultaneously, place the underside of the wax block on top of the warm metal knife or spatula and allow to remain until the wax melts and attaches itself to the holder.

Step 5.   Lift the wax block for a moment, withdraw the knife or spatula, and with a little pressure quickly place the block in position on the holder.

Step 6.   Pass a hot metal knife or spatula around all four edges of the base of the paraffin block to seal the block to the specimen holder.

Step 7. Allow the wax block to cool and harden (about 5 minutes). Since better cuttings are obtained when the microtome knife and wax block are cold, ice is often used to cool the knife and wax block before cutting.

CUTTING THE SECTIONS

The following procedure for section cutting with a rotary microtome can also be used with other types of microtomes with a few modifications, which depend on the design of the particular instrument.

Step 1. Become familiar with the feed mechanism, the adaptable knife holder, the specimen clamp, and the various adjustments of the microtome.

Step 2. Insert the block holder and specimen in the microtome clamp and tighten. The block and knife should be cooled before cutting. This is done by placing the block on a piece of ice for about 5 minutes, by allowing cold water to run on the tissue, or by leaving it in the refrigerator for 10 minutes. Cold specimens can be cut more advantageously, although prolonged icing causes the tissue to become soaked and hard to cut.

Step 3. Insert and tighten the knife in the knife holder. Make the necessary adjustments by elevating and tilting the knife so that it is in the correct position.

Step 4. Rotate the hand wheel slowly so that the wax specimen is brought to the level of the knife. Adjust and tighten the block holder by means of the adjusting screws, so that the specimen is held rigidly in the desired position.

Step 5. Adjust the cutting angle of the knife to approximately 6 to 10° from the vertical to produce good cutting sections.

Step 6. Set the cutting gauge at 20 micrometers, cut or trim the paraffin block until the tissue is exposed, and then cut a full tissue section. If a considerable amount of tissue is removed when a full section is cut, reset the paraffin block.

Step 7. Set the feed mechanism or thickness gauge to the desired tissue thickness. For routine work, sections 5 to 10 micrometers thick are adequate.

Step 8. Cut the wax section by rotating the handle of the micro-tome at a controlled speed to give good sections that are affixed to each other in the form of a straight ribbon.

Step 9. Support the ribbon sections as they come off the microtome with a spatula or flat instrument about 8 in. long. Stop the machine when sufficient sections are cut. Detach the last paraffin section adhering to the knife by means of a needle and gently lay the ribbon on a sheet of waxed or glazed paper spread flat on the table or on a hardwood board. It should be noted that the paper or board must be in a cool atmosphere free of dust and air currents. Straighten out the ribbon so that it lies flat on the table or hardwood surface.

Step 10. If required, cut additional sections in ribbon form and spread out alongside the previous ribbon.

Step 11. Separate the sections by means of a sharp scalpel or razor blade, using a forward stroke.

ATTACHING THE SECTIONS TO SLIDES—FLOTATION METHOD

The cut wax sections are slightly distorted and must therefore be flattened by warming in water heated to just under the melting point of the paraffin wax.

A tissue flotation bath (see Fig. 69) is used to heat the water quickly to the desired temperature. The temperature is kept constant by an internal thermostat.

Pour hot water nearly to fill a glass or metal receptacle measuring approximately 5 by 10 in. The temperature of the water should be 5 to 10° lower than the melting point of the embedding wax. Boil the water prior to use to eliminate entrapped air bubbles, which may attach themselves to the wax section.

Prepare a glass slide by placing a drop of an egg albumin mixture (see Appendix) or a substitute adhesive in the center of the glass slide. The egg albumin or adhesive is used to anchor or fasten the specimen

*Fig. 69. Tissue
flotation bath.*

to the slide.   Spread the adhesive evenly over the total surface of the slide, using a lint-free cloth such as nylon or acrylic fiber cloths, and then flood with water.

Transfer the specimen to the surface of the hot water bath by lifting the section carefully with a pointed needle, such as a plain or dissecting needle, being careful not to puncture the specimen. (Another method is to place the wax section on a clean slide and then transfer it from the slide to the bath by running dilute alcohol under the specimen on the slide, forcing the section to slip off the glass slide into the bath.)

Allow the section to flatten out in the bath.   Place the specially prepared albuminized slide vertically into the bath, touching one end of the specimen (the slide is immersed in the bath to a distance of about two thirds its length).   Withdraw the slide, still in the vertical position, and the specimen should adhere to the slide as it is withdrawn from the bath.   If the section does not adhere to the slide, use a dissecting needle to hold the specimen to the slide as it is removed vertically from the bath.

Place the slide on a flat surface and, with a dissecting needle, move the section to the desired position on the slide.

If a hot water bath is not available, the wax section can be transferred to the albuminized slide by the following alternate procedure: Transfer the wax section to the slide with a dissecting needle. Flood the slide gently with cold distilled water so that the wax section floats, but does not run off the slide. Heat the underside of the slide gently over a burner to a temperature just slightly below the melting point of the paraffin wax. This will cause the section to flatten out. Allow the slide to cool and drain off any remaining water on the slide; better still, blot the slide with a piece of lintless filter paper.

It is of the utmost importance at this stage that the specimen slide be thoroughly dried. During the drying procedure, the section flattens out thoroughly, wrinkles disappear, and the specimen adheres strongly to the albumin fixative on the slide. If the specimen is not dried thoroughly the section will float off the slide or loosen during the staining procedure. The slides are dried in a thermostatic oven for about 30 minutes at a temperature 10° below the melting point of the wax. An incubator kept at 37°C may also be used for drying wax sections. The slides are kept in the incubator overnight.

After drying, the slides can be stored in a slide box or in a dust-free atmosphere until ready for staining or for permanent mounting. It is a good technique to label and identify each slide by a serial number or other means.

REMOVING THE WAX FROM SECTIONS

It is essential to remove the paraffin wax completely from the section before the section can be stained adequately. To do this, warm the slide gently until the wax just melts or the temperature is just below the melting point of the wax (53°C). Soak the section in a Coplin jar containing xylene for 2 to 5 minutes or until the wax is dissolved. It is good procedure to repeat the immersion in a second jar containing fresh xylene. Remove the excess xylene from the specimen by immersing the slide in anhydrous alcohol. Repeat the immersion in alcohol (95%); then remove the alcohol by washing the slide several times in distilled water. If a mercuric chloride solution has been used for fixing the specimen, the mercury chloride deposits must also be removed from the section after the paraffin wax has been removed and before staining.

Place the slide in Lugol's iodine-potassium iodide solution (see

Appendix) until the specimen turns brown. About 5 minutes will usually suffice. Wash off the iodine solution in tap water; then wash the slide in a 5% sodium thiosulfate solution to remove the brown color. Wash the specimen thoroughly with water for about 1 minute.

Staining the Specimen

The microscopical specimen is stained so that any elements or cellular structures that are not readily visible under the microscope will become more prominent. Certain cell structures seem to have an affinity for specific dyes.

Many dyes and chemicals are used for routine and specialized staining. Each type of organic material, plant or animal, may require a somewhat different technique in staining to emphasize and to differentiate various parts. In the main, however, staining procedures are similar. In many cases, a mordant is used before adding the stain, so that certain tissue elements, such as the nucleus, will be selectively stained. Selectivity means that some of the cell structures will definitely take the stain while others will take it to a lesser degree or not at all. A mordant can be defined as a chemical used for fixing colors or dyes in substances by adsorption. Mordants are generally soluble salts of metals.

The major dyes employed in staining are acid or basic dyes; they generally resist being removed from the specimens when being washed with water or alcohol. Although specimens may be stained with one dye, staining the specimen with two dyes, one basic and the other acid, will stain many more cell structures than one dye and will also give greater color contrasts between different structures.

EQUIPMENT FOR STAINING

Jars of various sizes are necessary for containing the dye solutions. The type most often used for small staining operations is the Coplin jar (see Fig. 70), which are grooved on the bottom to hold about nine slides in a vertical position during the staining operation. The spacing of the grooves is such that the slides are positioned in a staggered pattern to allow the stain free access to the surface of the slides.

There are also staining jars, some with holders or racks, that hold

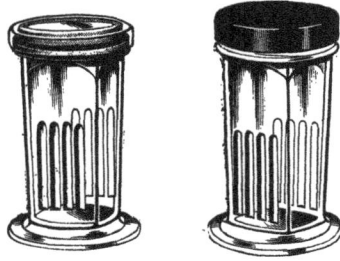

*Fig. 70.   Coplin
staining jars.*

many slides at one time (see Fig. 71).   The slides are stacked on the holder, which is then immersed in the jar of dye solution.

Fig. 71.   Staining jars and racks.

HEMATOXYLIN STAINS

Hematoxylin dye is the principal stain used in microscopy.   The pure dye, extracted from special wood trees such as logwood, consists of colorless or slightly yellow crystals, soluble in water, alcohol, and glycerin and slightly soluble in ether. In basic solutions, the color of the resulting stain is purple; in an acid medium, the color is yellow.

Before hematoxylin solution can be used as a staining reagent, the hematoxylin must be oxidized by exposure to air and sunlight. Oxidation can be accelerated by adding such oxidizing agents as sodium iodate.

When hematoxylin is used as a stain, it is essential that a mordant be used in conjunction with it. The mordant can be used separately, being applied first to the tissues and followed by the solution containing the hematoxylin dye, or an alum mordant is combined with the hematoxylin, the oxidizing agent, and a preservative in a single solution (see Appendix).

DIFFERENTIATION IN STAINING

In many cases, the hematoxylin is used purposely to overstain the specimen so that most of the structures are stained. The excess color is then gradually removed until the correct intensity of stain is achieved in the desired parts of the specimen. This widely practiced procedure of regressive staining is called *differentiation*. The general procedure is described below.

Step 1. Transfer the dewaxed specimen slide to the receptacle containing the hematoxylin stain. Allow to remain in the solution for about 20 minutes agitating the dye solution around the slides.

Step 2. Wash off the excess hematoxylin with a 1% solution of acidified alcohol. (The alcohol solution is prepared by adding 1 ml of concentrated hydrochloric acid to 99 ml of 70% ethyl or methyl alcohol.) The time necessary for washing must be determined empirically for each specimen. About 15 to 20 seconds will probably suffice for most specimens.

Step 3. Wash off the excess alcohol with water.

Step 4. Place the slide for 10 seconds in a 2% ammonium hydroxide solution to bring out the black or blue color of the hematoxylin dye.

(This process is called *blueing* the specimen. Precaution must be taken that the ammoniacal solution does not loosen the specimen and wash it off the slide.)

Step 5. Examine the specimen under the microscope and determine whether there is sufficient differentiation. If not, treat the slide again with a 1% acidic alcohol solution, as in step 3, and then repeat the subsequent steps. If, on the other hand, too much color has been washed out of the specimen, stain again with the hematoxylin solution for about 10 minutes and repeat the subsequent steps in the differentiation procedure.

PROCEDURE

Prepare the necessary stain solutions (see Appendix). Transfer them to individual Coplin jars.

Step 1. Transfer the dewaxed slide to a grooved Coplin jar containing the stain (Harris hematoxylin). Allow the slide, which is in a vertical position, to remain in the stain for 10 minutes or longer so that the specimen is overstained.

Step 2. Rinse the stained specimen in water.

Step 3. Differentiate the stain by immersing in a 1% acidic alcohol solution, prepared as described above, for approximately 5 seconds.

Step 4. Wash the slide in water for about 30 seconds to remove the excess acidic alcohol solution.

Step 5. Place the slide in a jar containing an alcoholic ammonium hydroxide solution (1 ml of ammonia water in 50 ml of 80% alcohol) for about 20 seconds. If at this stage the stained tissue section is examined under the microscope, it will be found that the nuclei of the cells are stained blue or black and are well defined, while the cytoplasm is almost colorless. As noted above, this step is called blueing the specimen. A 0.1% sodium carbonate solution may also be used for blueing.

Step 6. Rinse the slide in water for about 2 minutes.

Step 7. Counterstain the specimen by transferring to a Coplin jar containing an eosin stain (see Appendix) for about 2 minutes. (A counterstain is a dye that has little selectivity for the parts of the cell. It dyes everything indiscriminately to furnish a colored contrast or background for the discriminating hematoxylin stain. Other counterstains that can be used with hematoxylin stains are erythrosin, gold orange, fast green, and light green).

Step 8. Wash off the excess eosin with alcohol (95%).

Step 9. Transfer the slide to a jar containing absolute alcohol, allowing it to remain for 1 minute to dehydrate the specimen. It is essential that the slide be completely dehydrated before it is permanently mounted.

Step 10. Remove the alcohol by rinsing the slide with xylene for 2 minutes. Repeat the rinse with fresh xylene for another 2 minutes. The specimen is now ready for mounting.

Mounting the Specimen

If the specimen is to last for a long time, it must be sealed in a transparent liquid or solid medium. The liquid mountant chosen should harden quickly, not affect the color of the specimen stains, and dry to a smooth, hard finish without granulation.

CANADA BALSAM

The most widely used mountant is Canada balsam. It is a yellowish oleoresin liquid obtained from the buds of the balsam fir tree, found in Canada and the northeastern United States. The collected natural balsam is evaporated, and the solid is then dissolved and thinned to the right consistency with xylene. Canada balsam is sometimes referred to as *xylene balsam.* It is slightly acidic and is therefore neutralized by the manufacturer. When purchased, it is ready for use. The neutrality of the balsam can be maintained by adding a few grains of sodium carbonate to the balsam, which is kept in a special lightproof amber glass container called a *balsam bottle* (see Fig. 72). The balsam bottle is cylindrical and has a glass protecting cap ground to fit over the neck. It is also supplied with a glass applicator rod for applying the mountant to the specimen slide.

Canada balsam is advantageous in that it can be easily applied as a liquid and hardens rapidly to a transparent solid at ordinary temper-

Fig. 72. Balsam bottle.

atures.   It has approximately the same refractive index as the optical glass of the microscope, a property that is essential for good resolution. Other natural resins such as damar or sandarac may be used, but Canada balsam is superior in its working characteristics to these natural resins.

Synthetic resins are now slowly supplanting the natural resins. These synthetic resins, such as clarites and permounts, dissolved in toluene or xylene, are superior to the natural resins in that they are absolutely neutral. They are transparent, do not affect the color of the specimen stains, do not yellow with age, and can be applied easily.

In mounting the specimen, it is important that the mountant be of the right consistency.   If Canada balsam is used, it should be of a thin syrupy consistency, so that it flows readily from the glass applicator. If the balsam is too thick, it makes adjustment of the cover glass on the slide difficult, it darkens the specimen, and it gives an untidy appearance to the specimen.   If the balsam is too thin, air bubbles may replace the solvent as it evaporates.   Should the balsam be too thick, add a little xylene or benzene to thin it to the right consistency. If too much solvent has been added and the mountant becomes too thin, place the balsam in a warm atmosphere and allow the solvent to evaporate until the right consistency is achieved.

PROCEDURE

It is desirable that the specimen be placed in the center of the slide (for the method of finding the exact center, see page 112). Mounting is then done as follows:

Step 1. Remove the excess clearing agent (xylene), which was used to replace the alcohol in the stained specimen, by holding the slide vertically and allowing it to run off. Aid this process by shaking gently. Return the slide to the horizontal position and use a lint-free cloth to wipe the excess xylene from the slide. Work the cloth as near as possible to the specimen·

Step 2. Put two small drops of the mountant on the specimen. Do not stir the balsam before using. Two small drops are sufficient for most mounts, although thicker mounts require slightly more mountant. The correct amount can be determined with a little practice.

Step 3. Invert the slide with the specimen and mountant and place it gently on a cover glass so that no air bubbles are entrapped. Invert the slide again as soon as the mountant begins to spread out.

The cover glass must be thoroughly dry; otherwise, an opaque, milky mount may result. A good method for removing any possible moisture on the cover glass is to hold it with a pair of forceps and pass it through a low flame several times before applying it to the mount.

Step 4. Adjust the cover glass if necessary, using a glass rod or dissecting needle to push it into position so that it is directly over the specimen.

Step 5. Put the slide in an oven kept at a temperature of about 40°C for about 24 hours or until the mountant is hardened sufficiently (but not completely). Do not dry in direct sunlight, since the light may bleach the stain or darken the balsam.

Step 6. With a razor blade, cut and remove the excess mountant around the cover glass. With a cloth moistened with xylene, remove completely any remaining mountant on the slide near the cover glass.

Step 7. If air bubbles appear under the cover glass directly over the specimen, thus blocking the view of the specimen, soak the slide in xylene until the cover glass can be removed easily. Then wash the slide in absolute alcohol, rinse with xylene, and repeat the mounting procedure.

Celloidin Sections

Another method of preparing sections for cutting is to use celloidin instead of paraffin wax to impregnate the tissue.   The advantages of this procedure are that larger sections can be cut and less distortion occurs in the section because the celloidin, which holds the specimen structures permanently together, is not removed from the specimen, as is the paraffin wax.   Celloidin is mainly used to give support to specimens that are fragile, brittle, or friable.   This technique, however, has certain disadvantages.   The main objection is that the procedure may require a month or more.   It is very difficult to cut thin celloidin sections if the sections have not hardened properly; also, the celloidin procedure does not permit the cutting of sections in ribbons, so that each section must be cut separately. Finally, special stains that do not stain the hardened celloidin, but only the specimen proper, may have to be used.   If the nature of the specimen permits, the paraffin wax technique is preferred.

Celloidin, a concentrate of pyroxylin or nitrocellulose, is especially prepared for embedding specimens.   It can be bought in the dry, solid form, and a solution of known concentration can be made by dissolving the celloidin in a solution of ether and absolute alcohol. Celloidin solutions of 2, 4, 8, and 10% concentration are usually prepared; they vary from very thin to very thick.

To prepare an 8% celloidin solution, weigh 8 g of the dry shredded or chipped celloidin and place in a bottle equipped with a glass stopper.   Add 40 ml of absolute alcohol to the celloidin, stopper the bottle, and shake.   The alcohol does not dissolve the celloidin, but softens it.   Allow to stand for 48 hours and then add 60 ml of anhydrous ether and agitate.   Dissolving of the celloidin may take several days, with occasional agitation.   It can be seen that the ratio of the volume of alcohol to ether is 40 to 60.   *This alcohol-ether mixture is highly inflammable and toxic, and the dissolving of the celloidin must be done in a well-ventilated room or hood.*   Ethyl cellosolve, less highly inflammable and toxic than the ether-alcohol mixture, is a good solvent for the celloidin and may also be used.

The technique of obtaining and mounting celloidin sections follows a pattern generally similar to that described for the paraffin method. An outline of the necessary steps follows.

Step 1. Fix the specimen in a 10% formol-saline solution (see Appendix), allowing the specimen to remain in the solution for about 6 hours.

Step 2. Place the specimen in a vessel containing alcohol (80%) for about 10 hours.

Step 3. Place the specimen in a vessel containing alcohol (95%) for about 10 hours.

Step 4. Place the specimen in a vessel containing absolute alcohol for about 24 hours.

Step 5. Transfer the specimen to a vessel containing an alcohol-ether mixture (in a 4:6 ratio) and leave it for 12 to 24 hours.

Step 6. Immerse the specimen in a 2% celloidin solution in a wide-mouth, glass-stoppered bottle and leave it for 5 days.

Embedding the specimen in celloidin is best done by first treating it with a dilute celloidin solution and then with a series of celloidin solutions of gradually increasing concentration, as outlined below.

Step 7. Transfer the specimen to a wide-mouth, glass-stoppered bottle containing a 4% celloidin solution and leave it for 5 days.

Step 8. Repeat step 7, using an 8% celloidin solution, and allow the specimen to remain in the solution for 7 days.

Step 9. Prepare a sufficiently large paper mold to house the tissue loosely. Place the tissue in the paper boat with the surface to be cut facing upward.

Step 10. Pour a 10% celloidin solution over and around the specimen, covering it to a depth sufficient to allow for any shrinkage during the drying period.

Step 11. Place the paper mold and specimen on a flat piece of plate glass and set in a deep glass receptacle or bowl provided with a cover. Put 5 ml of ether in a small beaker, place the beaker in the glass receptacle, and cover. Allow the specimen to remain for 3 hours, so that any air bubbles present in the celloidin will be eliminated.

Step 12. Remove the beaker containing the ether from the receptacle and replace with a beaker containing 10 ml of chloroform, the vapors of which harden the' celloidin.   Allow the chloroform to remain until the celloidin hardens sufficiently so that the block will retain its shape under pressure for good cutting.   Generally, 24 to 48 hours will be long enough.

Step 13.   Trim the block with a knife or razor so that a small, even margin of celloidin is left around the specimen.

Step 14.   Mount the celloidin block on a deeply serrated block holder, made of wood or hard rubber, in the following way : Place the surface of the holder in an alcohol-ether mixture (4:6) for 10 minutes.   Immerse the bottom of the celloidin block in the same solution for 2 minutes.   Remove the block holder and apply a thick layer of the celloidin solution (10%) to the surface of the holder.   Place the celloidin block on the holder and press tightly and firmly together. Allow the base of the celloidin block to harden in the presence of chloroform vapors, as described in Step 12.   If sections are not required immediately, store the celloidin block in alcohol (70%).

Step 15.   Use a slide microtome for cutting the celloidin section. Fasten the carrier block tightly in the microtome clamp. Set the knife or razor at the proper oblique angle and pass the blade through the celloidin block in one long, continuous movement.

The surfaces of the knife and block must be kept continuously wet with alcohol (70%) during the entire section-cutting to eliminate irregularities in cutting.   A piece of absorbent cotton saturated in alcohol or a camelhair brush can be used to keep the block and knife wet.   Remove each section as it is cut with a brush moistened in alcohol, and transfer immediately to an alcohol solution (70%).   Celloidin sections are usually cut 8 micrometers or more in thickness.

Step 16.   Stain, differentiate the stain, and counterstain the celloidin section by the method described for paraffin wax sections.

Step 17.   Transfer the section to a slide and mount in Canada balsam as described on page 140.

The preparation of a celloidin block may be accelerated by placing the glass-stoppered bottles containing the different celloidin solutions in a well-ventilated oven maintained at 60°C and immersing the specimen in each solution from 1 to 2 hours. *Note, however, that alcohol and especially ether are highly volatile and most hazardous at higher temperatures.*

Frozen Sections

This method is most advantageous when rapid preparation of a section is required, as, for example, before or during an operation, when a quick diagnosis of a tissue section is desired. A specimen can be cut and stained in 10 or 15 minutes by this method.

Frozen sections are further advantageous in that very little shrinkage occurs in the specimen during its preparation and it is more easily stained. One of the principal advantages is that microchemical analyses can be performed on the untreated specimen. Frozen sections are also used when there is a need for preserving fatty materials in the specimen that would ordinarily be dissolved by the reagents used in the paraffin method.

This technique has several disadvantages, however. For one, a special freezing microtome with an attachment that allows cooling of the knife must be used (see Fig. 73). The freezing agent generally used is carbon dioxide. It is also rather difficult to obtain very thin sections. The minimum thickness that can be cut is 10 micrometers. Some distortion of the specimen may also occur during the freezing and cutting, and some parts may not have a natural appearance. It has also been found that some difficulty is encountered when staining specimens derived from frozen sections. Finally, this technique is not suitable for preparing serial sections.

The steps in preparing, staining, and mounting a specimen by this method are as follows :

Step 1. Fix the specimen (about 0.5 cm thick) by placing it in a test tube containing a 10% formol solution and boiling for 1 minute. Rinse in water.

Step 2. Spread a thin layer of Hamilton's freezing mixture (see Appendix) over the surface of the microtome stage. If Hamilton's

Fig. 73. Freezing microtome with knife-cooling attachment.

mixture is not available, use a saturated solution of cane sugar. Water by itself may form ice and damage the microtome knife.

Step 3. Position the specimen on the freezing stage of the microtome. Moisten the specimen with a few drops of water and freeze with carbon dioxide by slowly releasing the gas contained in a cylinder. Continue the carbon dioxide flow over the specimen with successive jets until the specimen is coated with a white snow film. Do not freeze too hard. The specimen should be chilled, not frozen solid. If desired, the specimen can be enveloped and supported in a semifluid gelatin solution containing 0.1% phenol as a preservative. The specimen surrounded by the supporting medium is then placed on the freezing stage.

Step 4. Cut the frozen section with a sliding microtome knife. If the section is frozen too hard the section coming off the knife will crumble and chip; therefore, allow it to thaw a little before cutting. If, on the other hand, the specimen is not frozen sufficiently, the knife will spread the tissue and injure it; to prevent such damage, apply a few more bursts of carbon dioxide to the specimen.

Step 5. Remove the section from the knife with a moistened finger and transfer it to a container of water. Immerse a clean slide smeared with egg albumin or other adhesive and float the specimen onto it.

Step 6. Remove the slide and specimen, drain off the excess water, and blot the slide almost to dryness with a piece of absorbent filter paper.

Step 7. Rinse or flood the section first with alcohol (95%) and then with absolute alcohol.

Step 8. Apply a 1% celloidin solution over the section, inclining the slide so that the solution will spread in a thin film. Drop some alcohol (95%) on the section to harden the celloidin.

Step 9. Stain in the manner described for paraffin sectioning. If speed is essential, each step in the staining and washing procedure must not exceed 1 minute.

Step 10. Mount in Canada balsam or clarite by the method described for paraffin sections.

MICROTOME-CRYOSTAT

The microtome-cryostat (see Fig. 74) is an instrument that automatically yields frozen sections rapidly and efficiently. Basically, the

Fig. 74. Microtome cryostat.

instrument is a microtome mounted at a 45° angle in a stainless steel cold chamber in which the temperature can be regulated from −10 to −30°C. The timing of the freezing period (0 to 3 minutes) can be controlled from the panel of the instrument.

The fresh tissue can be quickly frozen in a specimen holder mounted in the cold chamber. The holder is then transferred to the microtome. As the specimen is cut, it is held flat against the knife by an antiroll plate to prevent curling. The specimen is then transferred to a slide held by a suction pickup device.

CLEANING AND POLISHING, LABELING, AND STORING FINISHED SLIDES

Cleaning and Polishing

Slides must be cleansed of any excess mountant that may have exuded around the rim of the cover glass. The greater part of the excess mountant can be removed by scraping with a small, dull knife, taking care not to touch the cover glass writh the knife. A small lump of absorbent cotton saturated in alcohol (95%) is then used to remove the remainder of the excess balsam. Although xylene might be used more advantageously to dissolve the hardened, exuded mountant, it is generally not used because it is so strong a solvent that it may loosen the cover glass and cause it to fall off the slide. The dried slide is then dipped into a dilute soap solution or mild detergent to remove any dirt, grease, or stains from the rest of the slide. Finally, it is polished clean with a soft cloth.

Labeling

Slides are usually labeled temporarily during preparation to avoid mistakes as to their subject matter and for identification for eventual permanent labeling. Various materials may be used, such as glass-writing ink, glass-writing pencil, engraving instruments, or small numbered labels. These temporary labels are wiped or removed from the slides during the cleaning and polishing procedure. The slides are then permanently labeled with special gummed labels that are available in boxes or in booklet form (see Fig. 75). This system of

```
Scientific Name _____
Common Name _____
Locality { Region_____
         { Habitat_____
Date _____      No. _____
Collector _____
```

Fig. 75. Slide identi-
fication label.

permanent labeling has been accepted by most workers. The labels usually have colored borders. Before the labels are affixed to the slides, the appropriate information may be typewritten, printed, or written with a fine pen in waterproof India ink.

To affix a label firmly to a glass slide, moisten the upper surface of the label slightly. Then moisten the lower surface well and allow the label to flatten out for a moment. Apply the label firmly to the slide at the left of the cover glass and press firmly all around.

Some slides, as noted previously, have at one or both ends a ground surface that accepts waterproof India ink or other marking for labeling.

Storing

Finished slides should be stored in a cool, dry place away from direct lighting, which might bleach or discolor stains. Slides are best stored in trays (see Fig. 76) or small, grooved boxes that hold the slides upright

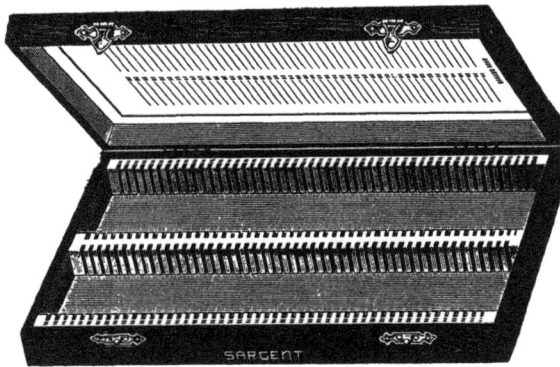

Fig. 76. Slide
storage tray.

and separated. They can also be stored flat in trays, but of course much more room is required with this method. For storing large numbers of slides, special storage cabinets with facilities for flat or vertical storage are available.

CHAPTER 6

Collecting and Preparing Pure Cultures of Various Organisms

PROTOZOANS

SOURCES FOR PROTOZOANS include quiescent ponds, rainwater that collects in small puddles in ditches or on grassy earth, and many others. Actually, a large variety of protozoans will be found any place where vegetation stagnates in water.

To collect samples, obtain several large wide-mouth jars with closures. Pass the mouth of the jar under the surface of water containing vegetation, such as grass, twigs, or leaves, or any soil. Fill the jar about 1/3 full with vegetation and water. Wipe the outside of the jar with a dry cloth, close the jar, and label as to origin of its contents.

Collecting can also be done with a plankton net (see Fig. 77), which is designed to retain organisms while straining out water. It is made

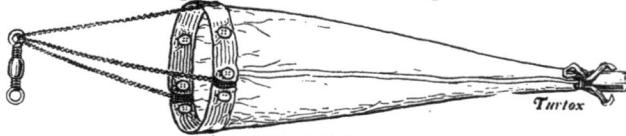

Fig. 77. Plankton net.

of silk bolting cloth, about 125 meshes to the inch, and is fastened to a heavy muslin ring, from which three braided leads converge to attach to a single towline. The net, attached to a rope or fishing pole, is cast on the water and slowly dragged back along the surface.   The opening of the net can then be closed by tying with a rope, or the organisms can be transferred to a wide-mouth jar with a cover.

Another common method of preparing cultures of protozoans is by means of infusions. To prepare an infusion, stock some dry hay, grass plants with roots attached, twigs, or lettuce leaves.   Cut this material into pieces from 1/2 to 3/4 inch long and transfer them to a large wide-mouth jar.   Do not fill the jar more than a quarter full.   Pour water into the jar within an inch of the top.   The water must be free from chlorine, which is generally present in drinking water.   Tap water can be boiled and cooled, or it can be vigorously stirred, allowed to stand for several days, and then used.   If distilled water is available, it will serve the purpose adequately.

Cover the jars loosely and place them near a window where they will receive diffused, but not direct, sunlight.   Allow to stand for several days. In this time, different varieties of microorganisms will develop in the infusion.   If specimen slides are prepared daily from the culture (from samples taken near the top and bottom and in the middle), it will be seen over a period of time that certain species appear for an interval and then disappear, while new species of microorganisms appear.

To prepare a pure culture of a given species of microorganism, it is necessary first to prepare a nutrient medium in which the microorganism will grow.   Such a medium is prepared by boiling some dry grass plants or hay in water for 20 minutes to kill most cysts of microorganisms and bacteria.   Allow to stand until the particles settle and then decant the supernatant liquid into several glass jars. Leave the jars open for about two days, or until a grayish scum appears on top of the liquid.   This nutrient medium serves as a base on which bacteria will develop to serve as food for the pure species of microorganism.

Pure Culture of Protozoans

Place a drop of the mixed protozoan culture on a slide and examine it under the microscope with the low-power objective. Make certain that the desired microorganisms are present on the slide. Place the tip of a capillary tube or pipet with a fine tip in the culture drop near where the organism is seen and draw up some of the liquid. Transfer this liquid to a fresh slide by blowing gently through the pipet.

Examine the slide again under low power. The slide will show the desired organisms in a more isolated condition. Repeat this procedure, each time using a fresh pipet and a new slide, until the desired organism is isolated when observed under the microscope.

Place the tip of a clean capillary tube in this droplet and gently draw up the single organism and then blow the contents into the prepared culture dish or jar. Repeat this process, transferring an isolated organism to each of several prepared culture dishes, so that if one does not produce the desired protozoan, one or more of the others will. Cover the receptacles loosely and put in a warm place. Within two days, many microorganisms will be reproduced from the single one.

In time the organisms in pure cultures may die out. To maintain the species over long periods, transfer some of the organisms from the original culture jar to fresh ones.

DIATOMS

Diatoms (see Fig. 78) are microscopic plants or algae. Among the most abundant organisms on earth, they are found in a great variety

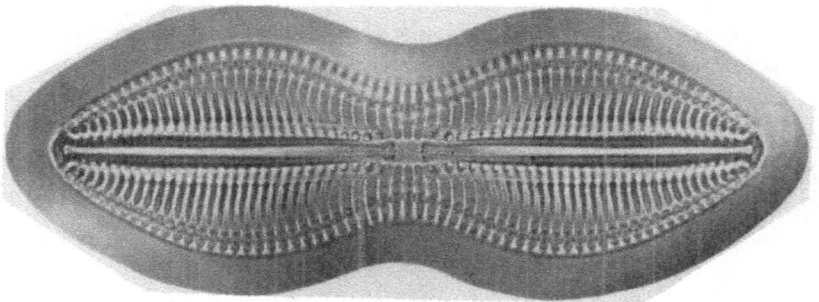

Fig. 78. Photomicrograph of a diatom (Navicula crabra).

of species. They occur in many different geometrical patterns, and exist largely in oceans and seas. Most of them are unicellular, with cell walls composed mainly of silicon dioxide. Since the cell walls of most diatoms are chemically inert, large deposits have accumulated over the centuries. Some of these deposits of fossil diatoms or diatomaceous earth were elevated out of the sea by the upheavals of the earth's surface. Occasionally, such deposits are almost 3,000 feet thick.

Diatoms can be found in early spring and autumn along the banks of standing bodies of water and on rocks and stones near these bodies of water as slimy masses or eelgrass. Diatomaceous materials scraped off the rocks or brown slimy masses collected from the surface of the water are placed in jars containing an equal amount of water.

The collected diatoms must be cleansed of sand, mud, and other debris before pure cultures are prepared. One method is to shake or agitate the diluted mixture. The very coarse materials, such as sand or organic materials, will settle to the bottom. The upper cloudy liquid is then poured into another jar, more water is added, and the mixture is shaken again. This procedure is repeated until most of the coarse material is removed and the remaining material is composed of diatoms, as determined by examination under the microscope.

Another method is to strain the diluted diatom mixture through a coarse cloth to remove the larger debris. The strained material is allowed to settle for an hour or two, after which the supernatant liquid is poured off and discarded and the fine sediment is allowed to remain on the bottom. Again add water to this sediment, shake, allow to settle for 2 hours, and pour off the supernatant liquid as before. Repeat this procedure until the water in the jar is clear and some suspended matter remains on the bottom.

After clearing, the jars are lightly covered, an equal amount of sea water is added to each, and they are then placed in diffused or subdued light at an optimum temperature of about 15 to 16°C for a period of time sufficient to allow an abundant growth of diatoms to occur. These diatoms can be used as a stock culture from which single diatoms can be removed and suspended in filtered seawater. Seawater appears to be the best medium for cultivating diatoms, since it contains specific salts.

Pure Culture of Diatoms

The method consists of preparing a special sterilized salt solution by adding certain nutrient salts to seawater and then sterilizing it (see Appendix for the preparation of Miquel's solution). This solution is then inoculated with a single species of diatom. The procedure for isolating a single species of diatom is the same as that described for isolating pure cultures of protozoan.

Another method (Allen and Nelson) of obtaining pure cultures of diatoms is to place one or two drops of the stock culture in 250 ml of the sterile medium (seawater and Miquel's salt solution in equal parts) in a flask. Shake the mixture and then pour into several Petri dishes. Place these dishes in subdued light at a constant temperature of about 15 to 16°C for several days. Examine the cultures, with a magnifying glass each day without disturbing them. Different species of diatoms will be seen growing separately on the bottom of the petri dishes. Do not permit culturing for too long a period, since this will allow one strong species to dominate. Select and pick up one distinct species and transfer it to a fresh sterile medium in a petri dish. This procedure can be repeated until the desired species is isolated.

BACTERIA

Given the right conditions of food, moisture, and temperature for growth, one bacterial organism can multiply in a short interval into a colony of similar bacterial organisms that are characteristic of the species. Such a colony of bacteria can be seen with the naked eye.

Pure Culture of Bacteria

Although certain bacteria may require different conditions, such as type of nutrient medium, temperature, and the presence or absence of oxygen, a general procedure can be applied for preparing a pure culture of bacteria.

The following equipment is essential : incubator, steam sterilizer, hot-air sterilizer, nichrome or platinum wire loop supported in a holder, needles, Petri dishes, and test tubes.

Transfer of bacteria or inoculation of a medium with bacteria is accomplished with a wire loop or inoculating needle. The loop or needle must be sterilized before use by heating in an open flame until red hot.   After use, the needle is again sterilized by heating gradually in a low flame to avoid any spattering or scattering from the wire or needle of any remaining organisms.

Prior to cultivating the bacterial organism, the culture medium must be prepared in test tubes or in Petri dishes.   There are many varieties of basic media that can be purchased and which are ready for use. It is also possible to prepare one's own medium. Nutrient agar composed of agar and beef extract is extensively used for this purpose.

STERILIZING GLASSWARE AND MEDIA

DRY-HEAT STERILIZATION

Since spores of bacteria may survive even when subjected to temperatures as high as 165°C, it is necessary to sterilize dry glassware or other equipment with dry heat in a hot-air sterilizer (see Fig. 79) at

Fig. 79.   Hot-air
sterilizer.

sufficiently high temperatures. The major factors in the destruction of bacteria are desiccation and coagulation. The best temperature for dry-heat sterilization ranges from 160 to 180°C. For complete sterilization, the duration at 160°C should be 1 hour; at 180°C, 10 minutes will suffice. The period of sterilization must be increased if more than a few pieces of glassware are put in the sterilizer at one time.

All glassware, such as test tubes and Petri dishes, must be clean and thoroughly dry before sterilization, because if sufficient moisture is present the temperature of sterilization may not go higher than 100°C.

To ensure that the glassware remains sterile, it should be put in suitable containers, which are then placed in the sterilizer.

MOIST-HEAT STERILIZATION

Nutrient medium for culturing bacteria must be sterilized in moist heat in a steam sterilizer (see Fig. 80). It is insufficient simply to boil the medium or expose it a single time to steam, either of which involves a temperature of only about 100°C, since many bacterial spores

Fig. 80. Steam
sterilizer.

with a low water content are not killed at 100°C.   It is possible to heat the nutrient medium repeatedly and thus in time weaken the spores so that sterilization can be achieved.

To really ensure sterility, however, the medium shall be heated under presure in an autoclave (see Fig. 81) or pressure cooker. Under

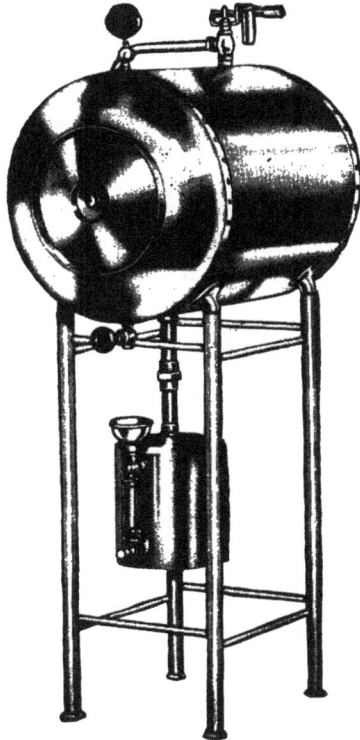

Fig. 81. Autoclave.

a pressure of 10 or 15 psi, the temperature reached will be about 115 to 120°C.   At this temperature, the medium can be effectively sterilized in 15 to 30 minutes.

Since media containing sugars may break down at autoclave temperatures, such media must be sterilized at lower pressures for longer periods. *Note that utmost care must be exercised in using an autoclave or a pressure cooker.*

Nutrient agar is commercially available in dehydrated form. The correct amount of water is added to the medium and the mixture is then steamed until the ingredients are dissolved. The pH of the mixture should be from 7.0 to 7.4, as determined either colorimetrically or electrometrically. If the pH of the medium is too low, adjust it by adding the necessary quantity of sodium hydroxide (0.05N). Steam the mixture again for 30 minutes and place appropriate portions in test tubes or Petri dishes, using a funnel to transfer the medium.

If many tubes must be filled at one time, a more efficient method of transferring the medium is required. In that case, place a large aspirator bottle on a flat table or stand and attach a piece of rubber tubing to the outlet tip of the bottle. Insert the glass part of a medicine dropper into the other end of the rubber tubing, fit the rubber tubing with a pinch clamp, and fill the bottle with liquid medium. To prepare a test tube, place the medicine dropper in the mouth, release the pinch clamp, allow the medium to flow into the test tube to the desired level, and then close the pinch clamp (see Fig. 82).

Plug the test tubes with sterile absorbent cotton. The cotton plug should be inserted, with the aid of a rod, about a quarter of the way down the tube, with a length of the cotton left protruding from the mouth of the tube for easy removal.

Sterilize the medium in an actoclave at 15 psi for 15 minutes.

Disposable Plugs

Disposable plugs of various types can be used instead of cotton. They can be readily manipulated and easily inserted without contaminating the tube or its contents. Several varieties are commercially available.

Disposable plugs are usually of white, nonabsorbent, foam-like plastic material. They are available to fit all sizes of culture tubes. The plugs can be autoclaved without losing their shape. They offer several advantages over cotton plugs—they are more uniform than cotton plugs, they do not stick to the fingers, they leave no deposits in cultures or media, and, finally, they do not obstruct visual examination.

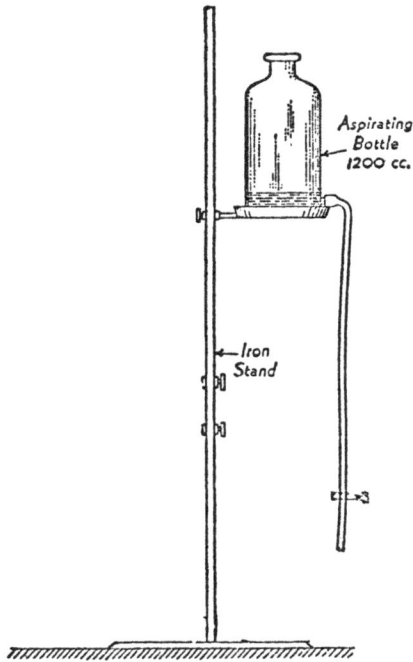

Fig. 82. Apparatus for transferring culture medium to test tubes (described in the text).

Disposable Sterile Culture Tubes

The use of disposable sterile culture tubes eliminates routine washing, sterilizing, and storage. They are generally made of biologically inert polystyrene, which can withstand prolonged heating at more than 80°C. Some have a double-action polyethylene cap that can be sealed for anaerobic cultures or left loosened for aerobic work.

PREPARING PURE BACTERIA CULTURES

Various methods are used to obtain pure cultures of bacteria. Two of the more common procedures are described below.

STREAKED-PLATE METHOD

Melt a tube of sterile nutrient agar in hot water and then cool to 40 to 50°C. Remove the plug from the tube and pass the tube opening through a flame. Lift the cover of a sterile Petri dish (see Fig. 83) by its edge, add 10 ml. of the agar, and allow to harden.

Fig. 83. Petri dish.

Slowly dip a loop that has been sterilized by flaming into the specimen or inoculum from which a pure culture is to be made. Then, starting at one edge of the agar dish, touch the loop to the surface and draw it back and forth in a zig-zag pattern across approximately one quarter of the surface of the agar plate. After completing this first streaking, lift the loop and sterilize it again slowly in a flame.

Place the sterilized loop in the inoculated agar again and, with a single   stroke, carry the specimen out from the   inoculated area. Repeat the zig-zag procedure over half of the remaining sterile area, at right angles to the original streaked area. Remove the loop and sterilize it slowly in a flame.

Place the sterile loop in the second inoculated area and, with a single stroke, carry the specimen out from the inoculated area. Repeat the zig-zag procedure over the remainder of the sterile area, at right angles to the second inoculated area. Remove the loop and flame slowly.

This specific procedure is followed so that the first portion of the inoculated agar will give massive growth, while the other sections will give more discrete and isolated bacterial colonies. If only massive growths of pure cultures are desired, inoculate each third of the agar surface with a fresh specimen. Place the inoculated plates in a bacteriological incubating oven (see Fig. 84) maintained at 37°C and allow to remain for 24 hours.

Fig. 84. *Bacteriological incubating oven.*

POURED-PLATE METHOD

Melt three tubes of agar (marked No. 1, 2, and 3); then lower their temperature to 45°C in a water bath. Remove the plug from the tube containing the bacterial specimen and pass the mouth of the tube through a flame. The withdrawn plug should be held in the fingers in such a manner that the end that is to be reinserted into the tube does not become contaminated. After passing the mouth of the tube through the flame, replace the plug into the tube. Repeat this procedure with the three tubes of agar.

Sterilize a nichrome or platinum wire with a needle or loop at its end by heating in a flame until red hot. Allow to cool, and remove the plug from the tube containing the bacterial specimen. Dip the wire into it to a depth of about 5 mm.

Unplug the No. 1 agar tube and insert the point or loop of the wire with the bacterial specimen into the agar. Rotate the wire several times. Remove the wire, pass the open mouth of this tube through a flame, and replug the tube.

Without redipping the wire and taking care not to contaminate it by touching it to anything, unplug tube No. 2, dip the wire into the agar, and rotate the wire as before. Remove the wire, pass the open mouth of this tube through a flame, and replug the tube.

Unplug tube No. 3 and follow the same procedure.

Finally, flame the wire slowly and carefully to sterilize it. Pass the open mouth of the tube containing the original bacterial specimen through a flame and replace the plug.

Obviously, tube No. 1 will have a heavier concentration of bacteria than No. 2, which in turn will have a heavier concentration of bacteria than tube No. 3.

Rotate the tubes of agar between the palms of the hand to mix the bacterial specimens with the agar. Remove the plug from tube No. 1, lift up the cover of a sterile Petri dish (also marked No. 1), and pour the contents of the tube into the dish. Replace the cover and, keeping the dish horizontal, gently agitate the agar by moving the dish in a circle. Leave the cover open slightly for a moment to allow the water vapor to escape. Replace the cover, allowing the dish to cool and the agar to solidify. Repeat the procedure for tubes No. 2 and 3, put the Petri dishes into an incubator maintained at 37°C. A period of 24 to 48 hours will suffice for most bacteria, although certain species may require several weeks of incubation.

ISOLATING PURE CULTURES FROM BACTERIAL COLONIES

After incubation, the Petri dishes are removed from the incubator and examined. It will be found that colonies of bacteria have grown on the surface of the agar; some of these colonies will be isolated, others with be lumped together (see Fig. 85). Since each discrete colony consists of many bacteria that have resulted from the reproduction of one bacterium, each colony will generally represent a pure culture.

To isolate such a pure culture of bacteria, examine the plate with a magnifying glass or low-power microscope and locate a discrete, well-isolated colony. The position of the colony can be marked on the bottom of the dish with a marking crayon. Flame an inoculating needle (not a loop) and allow it to cool. Touch the chosen colony precisely with

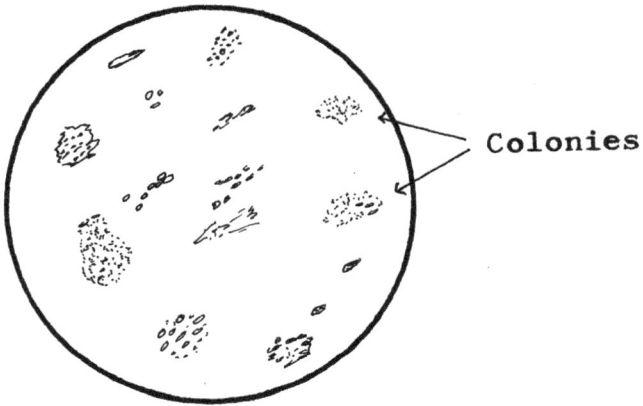

*Fig. 85. Petri dish showing growth of
bacterial colonies.*

the tip of the needle, carefully avoiding any other colony. Dip the
needle into a tube of sterile nutrient agar or other culture medium
and mix the contents. Incubate the tubes at 37°C, inclining them in
the incubator so that a greater surface area is available for bacterial
growth, thus assuring good growth of a pure bacteria on the agar
slants.

CHAPTER 7

The Polarizing Microscope

POLARIZED LIGHT IS USED in microscopy to bring out details and structures in transparent specimens for which ordinary transmitted light would not be effective. It is used extensively in the examination and study of thin sections of minerals, crystals, chemicals, ceramics, synthetic and natural fibers, and paper. It has proved valuable as an aid to research in geology, chemistry, and biochemistry. In biological microscopy, polarized light is invaluable for obtaining a more detailed picture of cell and tissue structure.

Although a standard microscope can be equipped with the necessary accessories to polarize light, special microscopes called *petrographic* or *chemical* microscopes are available with these accessories built in. They are equipped for quantitative measurement and generally finer work as compared to the conventional compound microscope with removable polarizing accessories.

The polarizing microscope is a conventional compound microscope with a few additional optical parts integrated into its lens system and with some change in its design to allow for quantitative work. The design of a polarizing microscope varies with the manufacturer, although some components are common to all.

The typical polarizing microscope (see Fig. 86) is equipped with a *rotating stage*, graduated in degrees, that can be rotated through 360°. The stage is provided with a vernier scale, which allows its position to be read to a fraction of a degree.

Instead of the conventional eye tube, the stage must be movable for

166

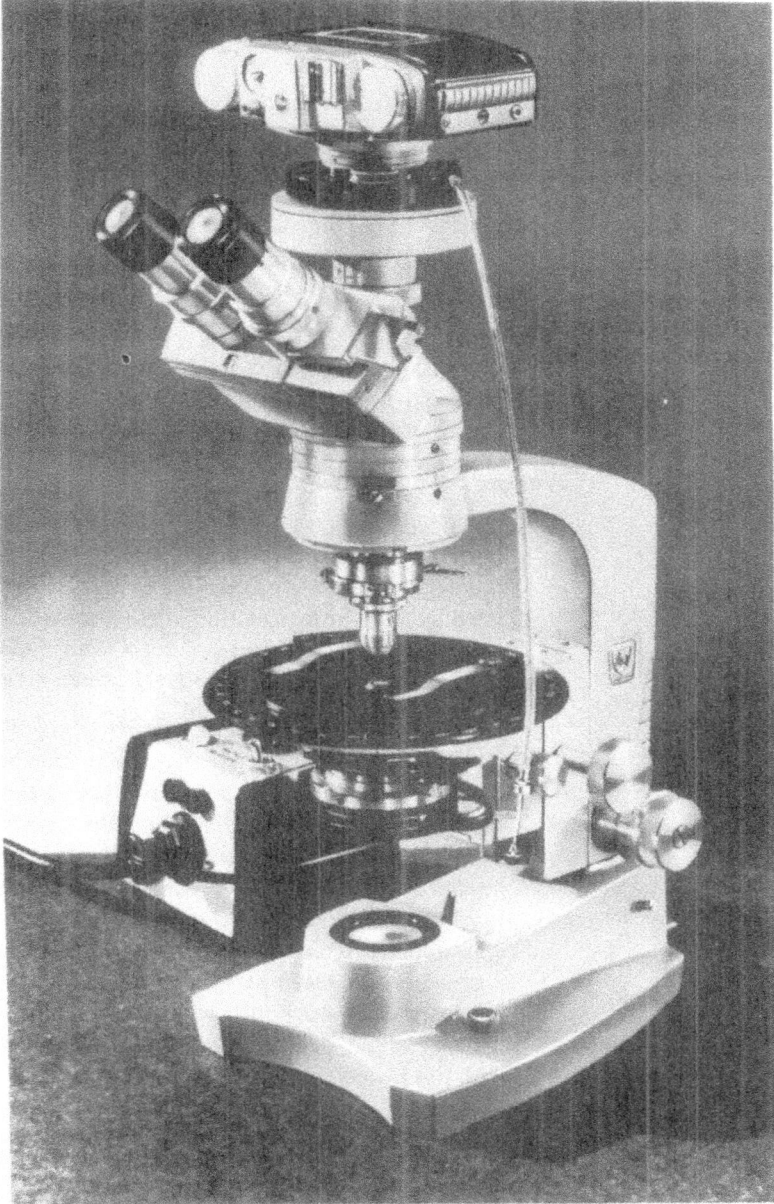

Fig. 86. Polarizing microscope.

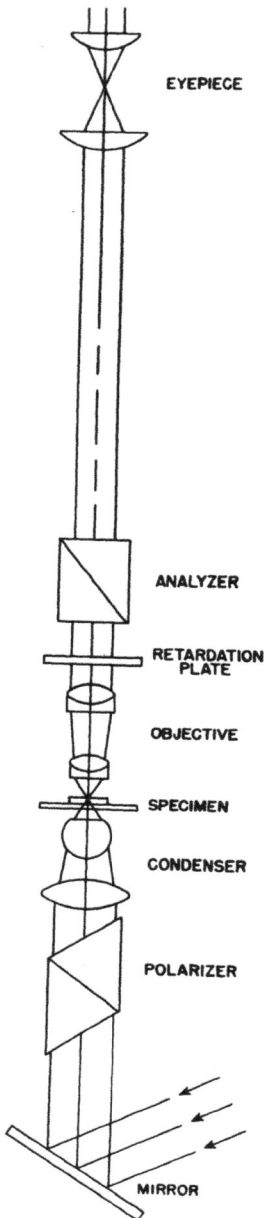

EYEPIECE

ANALYZER

RETARDATION
PLATE

OBJECTIVE

SPECIMEN

CONDENSER

POLARIZER

MIRROR

Fig. 87. Diagrammatic representation of the path of light through a polarizing microscope.

focusing, so that a source of illumination can be attached to the fixed eye tube when reflected, rather than transmitted, light is used for opaque specimens. With a movable tube, it would be necessary to realign the source of light each time the position of the eye tube was changed for focusing.

The *polarizer*, which is a calcite prism or a Polaroid plate, is mounted beneath the substage condenser or, in some polarizing microscopes, combined with the condenser as one unit. The polarizer may have a scale attached and can be rotated through 360° and set at any angle desired. Below the polarizer is an *auxiliary condenser*, which can be swung out of the way when not needed. Its function is to illuminate the field of view completely when the orthoscopic system is used.

The *analyzer*, which may be a calcite prism or a Polaroid plate similar to the polarizer, is mounted within the body tube above the objective. It can be rotated 180° in either direction. The analyzer is provided with a scale and vernier graduated in degrees. The analyzer, together with the polarizer, can be swung out of the optical path to permit conventional bright-field illumination. The path of light through a polarizing microscope as set up for polarization is illustrated in Fig. 87.

The *Amici-Bertrand lens* is situated between the eyepiece and the analyzer. This lens is used to obtain interference patterns of crystals, which aid in the identification of these crystals. It can be centered and focused on the back lens of the objective to bring the images of the interference figures into the focal plane of the eyepiece.   Some of these lenses are provided with an iris diaphragm to permit a variable field of view.   The lens can be moved out of the light path when required.

Beneath the analyzer and above the objective is a *slot* or receptacle that serves to mount accessories or compensators, which can be inserted or removed at will.   The slot is generally inclined at a 45° angle.

Special *Huygenian eyepieces* with internally mounted crosshairs are used.   They also have an adjustable eye lens, so that the crosshairs and the specimen under observation can be brought into simultaneous focus.   The eyepieces are generally wide-field types, which give a larger field of view than conventional eyepieces.

The special objectives supplied with polarizing microscopes are called *strain-free.*   Special care is taken by the manufacturer to fabricate and mount these objectives in such a way that no strain is developed within the objective, since any inherent strain might produce slight birefringence in the lens system or cause a slight rotation of the polarized light, which could lead to serious errors in measuring and evaluating specimens.   The objectives, corrected for use without a cover glass, are specially mounted in a nosepiece so that they can be optically centered.

ORTHOSCOPIC AND CONOSCOPIC ILLUMINATION

Two systems of illumination are generally used with the polarizing microscope to identify and classify substances—the *orthoscopic* and the *conoscopic* system (see Fig. 88).

To obtain orthoscopic illumination, the Bertrand lens is swung out of the optical path and the auxiliary condenser below the polarizer is swung into the optical path, so that the specimen is illuminated by parallel rays.   With this system, a true image of the specimen is seen.

To obtain conoscopic illumination, the Bertrand lens is swung into

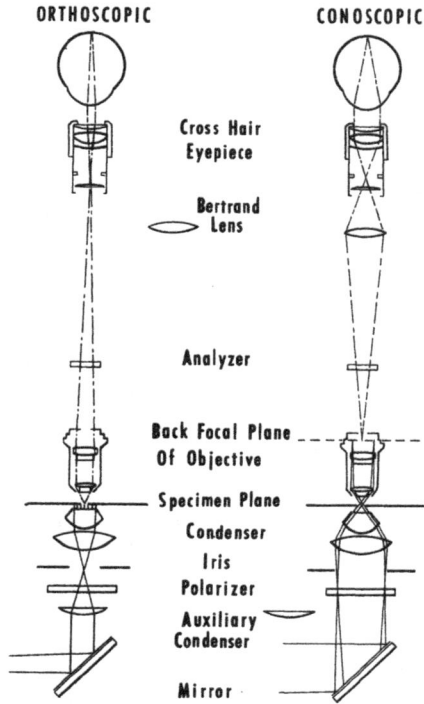

ORTHOSCOPIC   CONOSCOPIC

Cross Hair
Eyepiece

Bertrand
Lens

Analyzer

Back Focal Plane
Of Objective

Specimen Plane
Condenser
Iris
Polarizer
Auxiliary
Condenser

Mirror

Fig. 88. Diagrammatic representations of orthoscopic illumination (left) *and conoscopic illumination* (right).

the optical path and the auxiliary condenser is swung away from it, so that the specimen is illuminated by strong convergent rays from the substage condenser lens. The polarizing microscope is brought into focus on the back lens of the high-power objective, which then brings the entire optical system into focus at infinity. In this system, the Bertrand lens in fact acts as the objective, and the back lens of the objective becomes the viewed object.

NATURE OF LIGHT

In order to use the polarizing microscope properly, it is essential to understand some of the properties of light and its behavior.

Such an understanding is even more essential to proper use of the phase-contrast and the interference microscope, which are discussed in Chapters 8 and 9, respectively.

Theoretical Considerations

According to Maxwell's electromagnetic theory, light is a form of radiant energy that travels through space in continuous waves. These light waves consist of a narrow range of wavelengths in the electromagnetic spectrum, which consists not only of *visible* wavelengths of colored light (red, orange, yellow, green, blue, indigo, violet), but also of wavelengths that are not directly visible, such as the longer waves of infrared and the shorter waves of ultraviolet. These *invisible* radiations are included in the general definition of light.

The quantum theory, on the other hand, argues that light travels from place to place as discontinuous bundles of energy or particles called *quanta*.

Although there would appear to be a contradiction between a theory that describes the transmission of light in terms of waves and another that describes it in terms of particles, each theory is useful in explaining certain phenomena. Generally speaking, the emission and absorption of radiation by matter are explained in terms of quanta, and the transmission of radiation from one place to another in terms of electromagnetic waves or continuous wave motion.

Under ordinary circumstances, light travels in straight lines. The direction, which can be represented as a pencil of light or a straight line, is called a *ray*. The wave theory of light provides a more satisfactory explanation of certain aspects of the behavior of light, such as polarization, refraction, and reflection, but it is more convenient to describe and graphically represent the movement of light as rays that compose the beams. In the discussion that follows, rays will be used to give the varied positions of the wave front as it travels through space.

The wavelengths of various forms of radiation, including visible light, are listed in Table 2. These wavelengths are given in nanometers (1 nanometer$=1\times10^{-9}$ m, that is, 0.000,000,001 m).

The configuration, or geometric pattern, of light waves in motion

TABLE 2.   WAVELENGTHS OF VARIOUS TYPES OF RADIATION

| Type of Radiation | Wavelength (nanometers) | |
	Range	Representative Color
Hartzian (radio) waves	above 2.2×10^5	—
Infrared	$700–2.2 \times 10^5$	—
Visible spectrum	400–700	—
red	647–700	650
orange	585–647	600
yellow	575–585	580
green	491–575	520
blue	424–491	470
violet	400–424	410
Ultraviolet	10–400	—
X rays	0.01–10	—
Gamma rays	0.0005–0.14	—
Cosmic rays	below 0.00005	—

can be more easily understood from the following illustration of wave propagation: Assume that a rope is attached by one end to a wall and that a man is holding the other end in one hand.   If the man moves his hand up and down in a regular, continuous motion, a train of waves will be set in motion in the rope.   (Waves such as these, which recur at regular intervals, are called *periodic waves*.)   Light waves are similar to the transverse waves moving along the rope, the direction of the up-and-down motion being at right angles to the direction of propagation.   Such a wave has the following characteristics (see Fig. 89):   It travels, or oscillates, from crest to trough and back

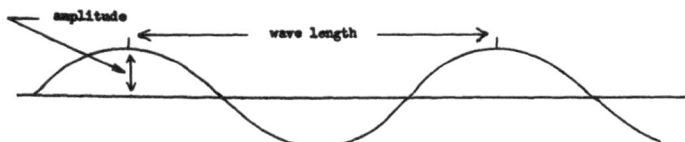

Fig. 89.   Diagrammatic representation of a light wave (discussed in the text).

in regular cycles, until the energy of propagation is dissipated.   The wave can be described as starting with a velocity of zero, proceeding to a maximum velocity, reversing itself, and returning to zero.   The *wavelength* is defined as the distance from the top of one crest to the top of the next; the *period,* as the time necessary to complete one vibration, from crest to trough; the *frequency,* as the number of wavelengths that pass a given point in one second; and the amplitude, as half the vertical distance from crest to trough.

Visible Spectrum

White light is not a single, simple form of radiation, but is in fact a blend of many colors of different wavelengths. It can be separated into its component colors by various methods, one common method being to pass a narrow beam of white light through a transparent glass prism, which bends, or refracts, the beam. Since light rays of different wavelengths are bent in varying degree, the various rays of light emerging from the prism, if allowed to fall on a white screen, spread out into a rainbow-colored spectrum, red at one end and violet at the other. The colors of the visible spectrum can be grouped broadly into six colors: red, orange, yellow, green, blue, and violet (see Fig. 90).

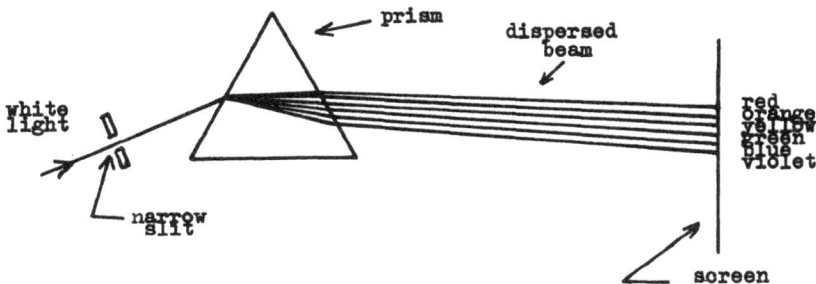

Fig. 90. *Diagrammatic representation of the dispersion of white light into the visible spectrum by a prism (discussed in the text).*

The wavelengths of the various colors have been measured, from the longer red portion of the spectrum to the shorter violet portion. It has also been observed that light of shorter wavelength is bent or refracted to a greater degree than that of longer wavelength. The colors that emerge from the glass prism blend gradually into one another, with no sharp dividing lines. Light of a particular wavelength, or of a very narrow range of wavelengths, is therefore described as *monochromatic light*.

The three most important characteristics of light in the visible spectrum are speed, wavelength, and frequency. The relationship among these quantities can be expressed mathematically by the equation $c = f\lambda$, in which c is the speed of light, a constant (in a vacuum, $c = 2.997925 \times 10^8$ m/sec, which is approximately 186,000 miles/sec); f is the frequency or vibration rate in cycles or periods per second; and λ is the wavelength of the light in micrometers, nanometers, or Ångstrom units (1 Ångstrom unit, abbreviated Å, $= 1 \times 10^{-10}$ m).

Reflection

Reflection of light is its rebounding from a surface into the original medium in which it had been traveling. For example, in Fig. 91, a

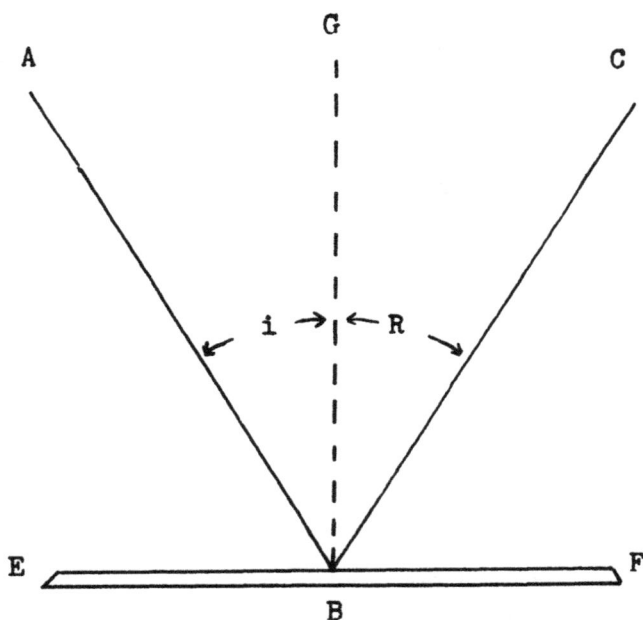

Fig. 91. Diagrammatic representation of reflection of light (discussed in the text).

ray of light, *A*, called the *incident ray*, falls on a plane surface or mirror, *EF*, at point *B*. The ray leaving the mirror, *C*, is called the *reflected ray*. An imaginary line, *GD*, perpendicular to the mirror surface, *EF*, is called the *perpendicular* or *normal*. It is a basic principle of reflection that regardless of the angle at which light strikes the reflecting surface, the *angle of incidence, i* (that is, the angle between the incident ray and the perpendicular), is equal to the *angle of reflectance, R* (that is, the angle between the reflected ray and the perpendicular), both angles being in the same plane.

The extent to which light will be reflected from any surface depends on the nature of the surface. The reflection of light from a flat, smooth, or highly polished surface (such as a mirror) is called *regular* or *specular reflection,* since the rays of light leave the reflecting surface in the same relative order, without distortion, as the rays striking the surface (see Fig. 92, *left*). To the eye, the reflected image will appear

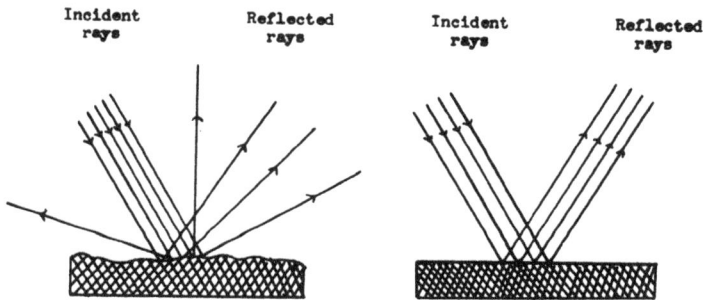

Fig. 92. *Diagrammatic representations of regular* (left) *and diffuse* (right) *reflection of light (discussed in the text).*

identical to the original image. Moreover, the reflected image appears to originate behind or beyond, rather than at, the mirror surface.

Light reflected from a rough surface, on the other hand, will be reflected equally in all directions; such reflection is therefore called *diffuse reflection* (see Fig. 92, *right*). Most dark objects that are not smooth become visible when they are struck by light. Furthermore, the object itself appears to be the source of light, since the actual light source is not seen, as is the case when light comes from a polished reflecting surface. A dark object is therefore visible only by the light that it diffuses. The intensity of the reflected light is less than that of the incident light because of the partial penetration of the light into the struck surface.

Refraction and Index of Refraction

Light ordinarily travels in straight lines as long as it continues to travel in the same medium. For example, light traveling through air travels in a straight line. When a beam of light passes from one medium into another, as from air into water, the velocity or speed of the light is changed; as a consequence, the ray of light seems to bend. This phenomenon is called *refraction*.

A common example of refraction is illustrated by a rod in a glass of water. The rod appears to be broken at the surface of the water, and below the surface it appears distorted.

An imaginary line, *n*, called the *normal*, can be drawn perpendicular to the plane of intersection between the original medium and the new medium at the point where the light enters the new medium (see Fig. 93). A ray of light entering a new medium that is denser

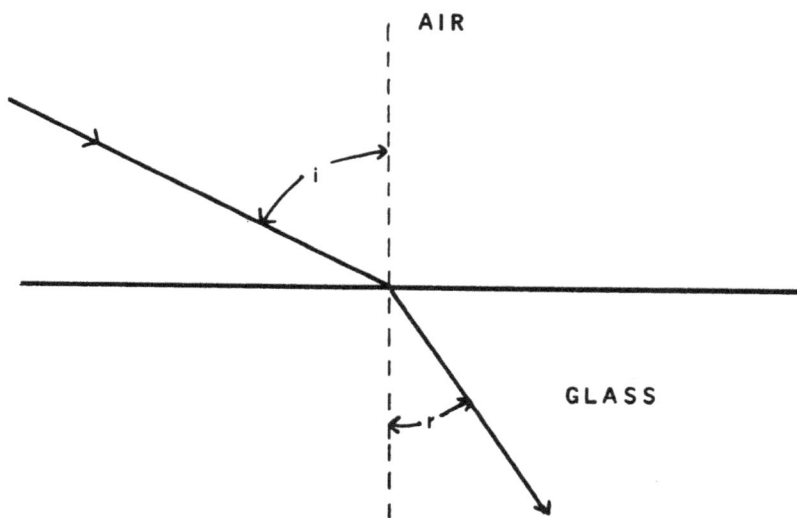

AIR

GLASS

Fig. 93. Diagrammatic representation of refraction of light (discussed in the text).

than the original will be bent toward the normal because of the reduction in the speed of the light. The denser this new medium, the greater the bend of the light toward the normal. On the other hand, if the light passes from a dense medium to a rarer medium, it will be bent away from the normal because of the increase in the speed of the light.

The relationship between the angle of incidence, i, and the angle of refraction, r, with respect to the normal has been expressed as an *index of refraction,* defined mathematically by the equation:

$$\text{Index of refraction} = \frac{\text{sine of angle of incidence}}{\text{sine of angle of refraction}} = \frac{V_1}{V_2}$$

in which V_1 is the velocity of light in the first medium and V_2 is the velocity of light in the second medium.

All transparent substances (that is, substances through which light passes) have a specific index of refraction. The greater the density of the substance, the greater the index of refraction and, consequently, the greater the bending of the light. The degree of the refraction of the light depends on the angle at which the light strikes the substance and the density of this substance as compared to that of the medium from which the light is coming.

The index of refraction, or refractive index, of a substance varies slightly with the wavelength of the light used. The index of refraction of yellow monochromatic light in a vacuum is usually taken as unity. The index of refraction of air at ordinary temperature is 1.0002918, which is close enough to unity for practical purposes and is therefore taken as the standard.

Thus, the index of refraction for a given substance is also the ratio between the speed of light in air and the speed of light in that substance. A substance with a lower index will allow the light to travel through it at a greater speed than a substance with a higher index.

Critical Angle and Total Reflection

Light traveling from one medium and falling obliquely on a transparent object is partially reflected as it strikes the surface of this object. The remainder of the light enters the object and is bent toward or away from the normal, depending on whether the object is denser or rarer than the original medium.

For example, when the light travels from water to air—that is, from a denser medium to a rarer medium—the ray will be bent away from the perpendicular. As the angle of incidence in the water becomes greater—or, in other words, as the incident ray is being transmitted at a greater angle—the rays that leave the water to enter the air are bent closer and closer to the surface of the water. A point will be reached at which the angle of incidence is such that the emergent rays are exactly in coincidence with the surface of the water. This angle of incidence, called the *critical angle* (see Fig. 94), is considered the

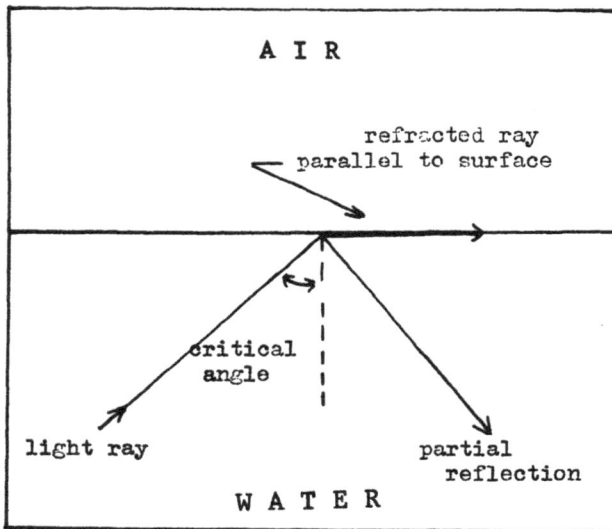

Fig. 94. Diagrammatic representation of critical angle (discussed in the text).

maximum angle of incidence at which refraction may occur when light is traveling from a denser to a rarer medium.

If the angle of incidence is greater than the critical angle, none of the rays will emerge from the water into the air. Rather, all the light will be reflected back into the water.   This phenomenon is called *total reflection* (see Fig. 95).

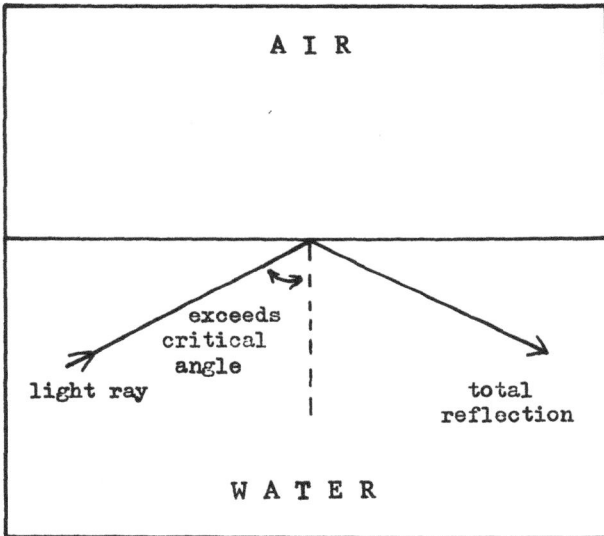

Fig. 95.   Diagrammatic representation of total reflection of light (discussed in the text).

The denser the medium from which the ray of light travels (as, for example, a dense crystal), the smaller the critical angle and, as a result, the greater the likelihood that total reflection will occur.   In crystals such as the diamond, which has a critical angle of about 23°, the rays of light may be totally reflected within the crystal several times before emerging into the air.

Polarization

Normally, waves of light traveling away from the source vibrate in all directions at right angles to the direction of travel or propagation. To put it simply, the waves of light are scattered and traveling in many directions or planes.   It is possible, by using such optical polarizing devices as the Nicol prism, Polaroid film, Ahren's prism, or birefringent crystals, to restrict the transverse light waves so that they vibrate in one specific plane or direction.   These selected light waves are called *polarized light.*

In polarization, light striking the polarizer prism is actually split into two polarized rays, termed the *ordinary ray* and the *extraordinary ray* (see Fig. 96), each of which constitutes half the original light. These two rays vibrate at right angles to each other.

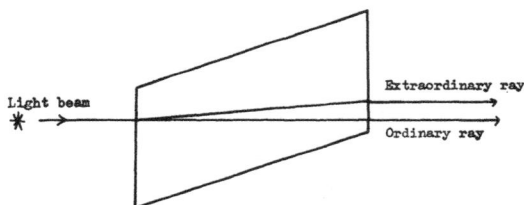

Fig. 96. Diagrammatic representation of the separation of a light beam in a crystal (discussed in the text).

In the polarizing microscope, an optical polarizing device, usually a pair of Nicol prisms or Polaroid film, is inserted in the path of the light beam. The ordinary ray is totally reflected within this device, so that it passes out of the field of the microscope and does not strike the eye; the extraordinary ray, on the other hand, continues straight through the optical axis of the microscope (see Fig. 97). The

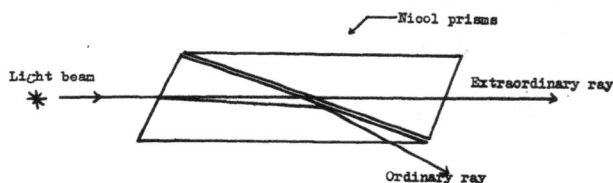

Fig. 97. Diagrammatic representation of total reflection of the ordinary ray (discussed in the text).

light thus transmitted through the microscope—that is, the extraordinary ray—is said to be *plane-polarized,* and the device itself is called the *polarizer.* It is usually positioned in a holder below the substage condenser.

If a second, similar polarizing prism or Polaroid plate, called the *analyzer,* is placed in the path of the polarized light coming from the polarizer, changes in the intensity of the illumination seem to occur as the analyzer is rotated. The analyzer is generally placed below the eyepiece or, if in the form of a round cap, over the eyepiece.

As one looks through a microscope equipped with these two polarizing accessories, it can be observed that as the analyzer is rotated, the light is diminished until at some point of rotation it disappears. At this point, the optic axis of the polarizer is at right angles to the optic axis of the analyzer. This relative positioning of polarizer and analyzer, in which no light reaches the eye, is called *crossed polarizers* (see Fig. 98, *bottom*).

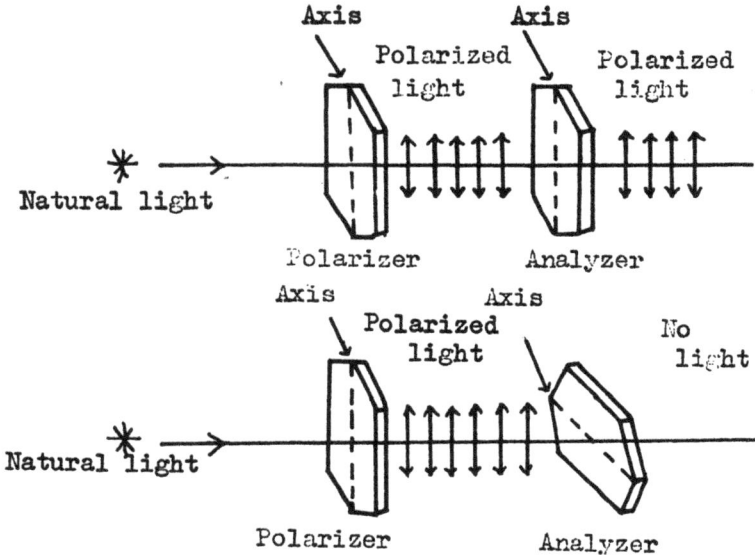

Fig. 98. Diagrammatic representation of polarized light: (top) *uncrossed polarizers* ; (bottom) *crossed polarizers.*

If a transparent crystalline (anisotropic) substance is placed on the microscope stage with the polarizers crossed, it will appear visible and bright when viewed through the eyepiece; when it is removed, however, the light will again disappear. This phenomenon results from the property of certain crystals or substances of interfering with the natural path of polarized light by altering, or rotating, its plane. What happens is that the polarizer and analyzer, originally crossed, are "uncrossed." Crystals or substances that rotate the plane of polarized light in this manner are said to be *optically active.*

Interference and Reinforcement

Light waves passing through a transparent substance such as a crystal and traveling along parallel paths may be partially reduced or retarded in velocity. That is, some of the light waves may encounter a thicker or denser portion of the crystal and therefore be slowed in comparison with the other waves. As light waves simultaneously reach a given point after leaving the crystal, a variation in the intensity of the illumination may occur. This variation in intensity is caused by the *interference* of the retarded light waves. Such interference may be *constructive,* in which case the illumination will appear brighter because of a reinforcement of the light waves, or it may be *destructive,* in which case the illumination will appear dimmed or darkened.

Thus, the degree of illumination afforded by light arriving at a given point depends on the wavelength, the amplitude, and the phase difference or agreements of these light waves. Two waves are said to be *in phase* when their amplitudes are in step-by-step correspondence, so that when they arrive at a given point there is constructive interference and they reinforce each other (see Fig. 99). Conversely,

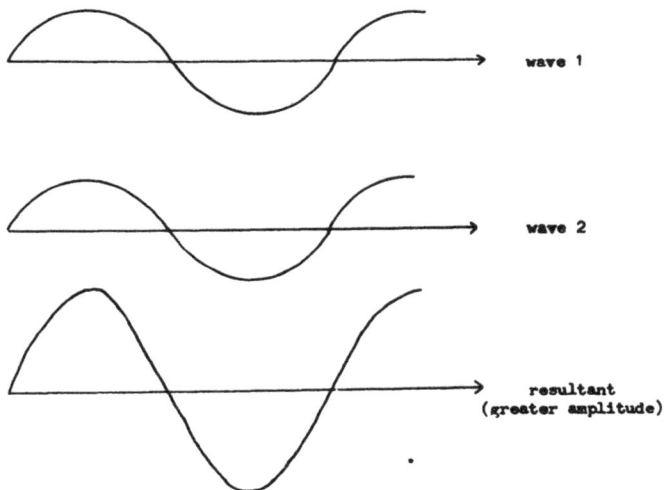

Fig. 99. Diagrammatic representation of constructive interference of light (discussed in the text).

two waves are said to be *out of phase* when their amplitudes are not in step-by-step correspondence, so that when they arrive at a given point there is destructive interference and they cancel each other partially or completely (see Fig. 100). The difference in phase, measured in wavelengths, is called the *path difference* of the light waves.

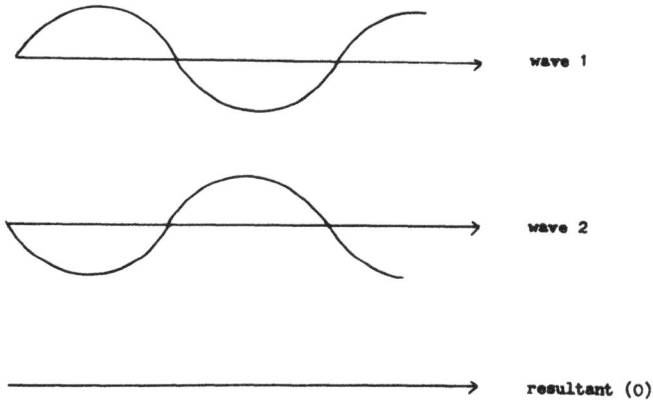

Fig 100. Diagrammatic representation of destructive interference of light (discussed in the text).

The phenomenon of interference can be illustrated more fully by an exercise in which a quartz wedge is placed on the stage of a polarizing microscope in a 45° position. The polarizer and analyzer are set in the crossed position. The specimen is illuminated with yellow monochromatic light (wavelength 580 nanometers), or a yellow filter is placed in front of the microscope lamp. As the specimen is observed through the microscope, no yellow light is transmitted through the thinnest edge of the quartz wedge, and this edge appears dark. The reason is that the crystal, being very thin at this point, does not produce any path difference in the light waves; therefore, since the specimen is observed under crossed polarizers, the thin edge appears dark (see Fig. 101, point *A*).

The wedge is now moved slowly under the microscope. As a slightly thicker portion comes into the field of view, interference occurs because of the refraction of part of the light. The path difference is such that this interference is constructive, and a yellow band of light is

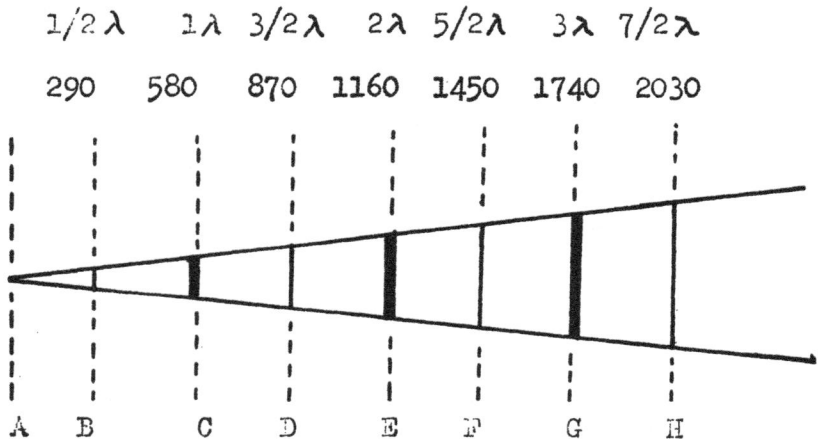

Fig. 101. Diagrammatic representation of interference of light by a quartz wedge (discussed in the text).

seen, although the specimen is still being observed under crossed polarizers. The path difference at this point is equal to half the wavelength of the yellow monochromatic light, or 290 nanometers (see Fig. 101, point *B*).

The wedge is again moved slowly until a slightly thicker portion comes into the field of view. Once again, there is interference of the light because of a path difference. The path difference now, however, is 580 nanometers, which is exactly one wavelength of yellow light. The waves at this point are out of phase and cancel each other, the result being that there is no illumination and the specimen appears dark (see Fig. 101, point *C*).

As the wedge is again moved slowly under the polarizing microscope, the changing thickness of the wedge produces a series of alternating black and yellow bands. It can be seen from Fig. 101 that if the path difference is equal to half the wavelength of the yellow light or any odd multiple thereof (that is, 580/2, $3 \times 580/2$, $5 \times 580/2$, $7 \times 580/2$,...), there is illumination because the resulting interference is constructive. Conversely, if the path difference is one wavelength or any multiple thereof (that is, 580, 2×580, 3×580,...), there is darkness because the resulting interference is destructive.

If red monochromatic light is substituted for the yellow light, it will be observed that the series of black bands, or interference fringes, will be shifted farther along the length of the quartz wedge. This result demonstrates that the red-colored light has a longer wavelength than the yellow light and thus comes to a focus at a further point.

Note that the analyzer introduces an extra path difference equal to half a wavelength, so that destructive interference normally occurs when the phase difference between two waves of light having a common origin is equal to half the wavelength or any odd multiple thereof. Conversely, the waves are reinforced—that is, there is constructive interference—when the phase difference is equal to one wavelength or any multiple thereof.

If this exercise is repeated with white light, which is composed of many colors, instead of monochromatic light, a striking array of different and changing colors is observed. The white light passing through the crystal is resolved into its component colors, or spectrum. At some points on the crystal, destructive interference of at least one wavelength present in the white light may occur, and the color with this wavelength may be canceled out. Consequently, the remaining light that reaches the eye will be minus the color represented by the obstructed wavelength, and the light as a whole will appear colored. It is for this reason that anisotropic crystals show varied characteristic colors when viewed under a polarizing microscope and illuminated with white light.

CRYSTALLOGRAPHY

The polarizing microscope is an invaluable tool in the examination and identification of crystals and their structure. Although crystals assume a wide variety of geometric forms in nature, it is possible to classify them according to the way their faces crystallize. These faces, referred to as *axes,* are imaginary lines or directions; they help in describing crystalline structure. All crystals can be categorized in the following six *systems of crystallization* (the identifying capital letters correspond to those in Fig. 102):

A. Regular (or Isometric). The crystal has three axes of equal length. All three axes intersect at right angles.

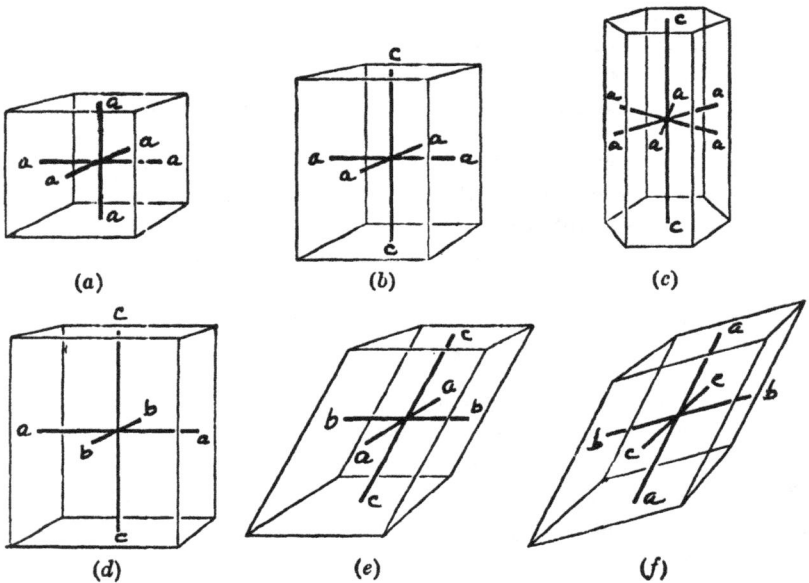

Fig. 102. Systems of crystallization (described in the text): (A) regular (or isometric); (B) tetragonal; (C) hexagonal; (D) rhombic; (E) monoclinic; (F) triclinic.

B. *Tetragonal.* The crystal has three axes, two of equal length and the third either shorter or longer. All three axes intersect at right angles.

C. *Hexagonal.* The crystal has four axes, three of equal length, in the same plane, and intersecting at 60° angles; the fourth axis, which may be either shorter or longer, is perpendicular to the plane of the other three.

D. *Rhombic.* The crystal has three axes of unequal length. All three axes intersect at right angles.

E. *Monoclinic.* The crystal has three axes of unequal length. Two of these intersect at right angles; the third is perpendicular to one of these, but not to the other.

F. *Triclinic.* The crystal has three axes of unequal length, which do not intersect at right angles.

Generally, when a crystal is to be identified in terms of its system of crystallization (the angle between any two given crystal faces is always the same for the same crystal), the axes and their point of intersection are plotted on graph paper. Since the crystal is three-dimensional, there will be three axes or coordinates (X, Y, Z) on the graph emanating from one point. If the X, Y, and Z axes are of different lengths, they are termed the a, b, and c axes, respectively.

Properties of Crystals

ISOTROPIC CRYSTALS

Crystals exhibit different properties depending on the position in which they are examined—that is, on the direction in which light is transmitted through them. An exception are crystals belonging to the regular (or isometric) system, which are found to exhibit the same values regardless of the position in which they are examined. In other words, monochromatic light travels through such crystals at equal velocity in all directions. Crystals showing these characteristics are called *isotropic*. When the polarizer and analyzer are crossed in a polarizing microscope, an isotropic crystal shows no optical properties when rotated on the microscope stage.

ANISOTROPIC CRYSTALS

Crystals in the remaining five systems of crystallization are considered *anisotropic* crystals. Their properties depend on the direction or position in which they are examined. Light travels at a variable velocity through such a crystal, depending on the direction of transmission of monochromatic light through the crystal. An anisotropic crystal appears alternately dark and light as it is rotated on the stage of a polarizing microscope with the polarizers crossed.

UNIAXIAL CRYSTALS

Uniaxial crystals, which belong to the tetragonal and hexagonal systems, have the following characteristics: When they are examined under polarized light, there will be *one* direction in which all monochromatic light travels at equal velocity. This direction is called the *optic*, or *C axis*; the existence of only one such axis in these crystals gives rise to their designation as uniaxial. If these crystals are examined with transmitted light in other directions not along the optic axis, it will be found that the beam of light is split into two different components with different velocities and indices of refraction.

Biaxial crystals belong to the rhombic, monoclinic, and triclinic systems. Since they have two optic axes, they are more complicated in structure than uniaxial crystals. When biaxial crystals are examined under the polarizing microscope, there will be two directions in which monochromatic light travels at equal velocity. Additionally, biaxial crystals are characterized by having three indices of refraction for light transmitted through them in any direction other than along the two optic axes.

BIREFRINGENCE

The phenomenon of *birefringence*, or *double refraction*, is characteristic of uniaxial crystals. It is also characteristic of liquid crystal substances. When a beam of light enters a crystal having the property of birefringence, the beam is split into two components, the transverse vibrations of which are at right angles (see Fig. 103).

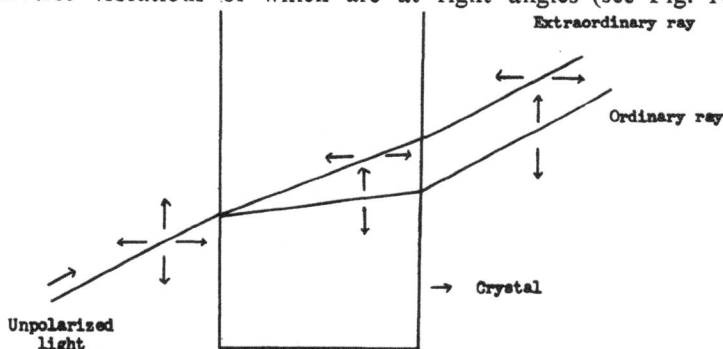

Fig. 103. *Diagrammatic representation of birefringence, or double refraction (discussed in the text).*

These two rays travel at different velocities and therefore have different indices of refraction—that is, they are refracted at different angles—as they pass through the crystal. The two rays emerge from the crystal as parallel rays of polarized light, one vibrating parallel, the other at right angles, to the optic axis.

One ray, called the *ordinary ray,* passes through the crystal without deviation. This ray travels through the crystal at the same velocity in any direction and obeys the laws of refraction. The other ray, called the *extraordinary ray,* deviates as it emerges from the crystal

This ray travels at a different velocity, and its behavior is dependent on its direction through the crystal.

If the two rays travel along the optic axis, or grain, of the crystal, their speed is the same and they coincide. If the extraordinary ray is traveling in another direction, however, its speed may be greater or less than that of the ordinary ray and will be greatest when the ray is traveling at right angles to the optic axis. If the maximum refractive index of the extraordinary ray is greater than that of the ordinary ray, the crystal is said to be *optically positive*. If the reverse is true, the crystal is said to be *optically negative*.

Birefringence can be demonstrated by placing a calcite crystal over a single dot marked on a piece of white paper (see Fig. 104). As one

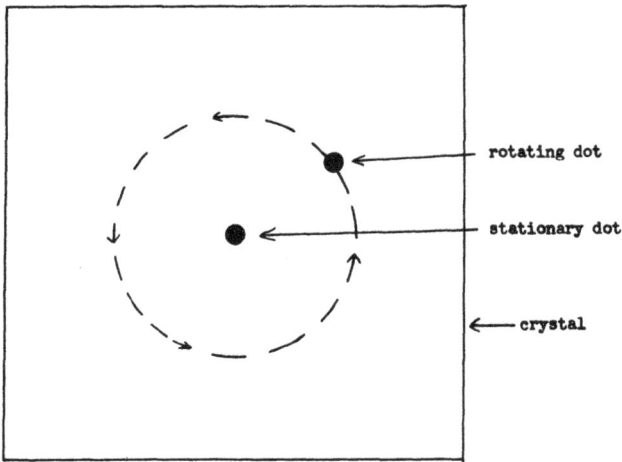

Fig. 104. Demonstration of birefringence in a calcite crystal (discussed in the text).

looks through the crystal at the dot, two dots will be seen. If the crystal is rotated, it will be observed that one dot remains stationary while the other dot rotates around it.   The ray of light coming to the eye from the stationary dot is the ordinary ray; that coming from the other dot is the extraordinary ray.   It is possible to measure the refractive index of each ray, and the difference between these two indices will give an indication as to whether the crystal is strongly or weakly birefringent. Thus, it is possible to identify a crystal by determining its degree of birefringence.

For example, quartz is a uniaxial birefringent crystal. The ordinary ray, the fast direction, has a refractive index of 1.54442. The extraordinary ray, the slow direction, has an index of 1.5533. The numerical difference between the maximum and minimum refractive indices is 0.0091. This difference is very small, and quartz shows weak birefringence.

Calcite, however, is a uniaxial crystal that shows strong birefringence. The refractive index of the ordinary ray is 1.6585; that of the extraordinary ray is 1.4864. The numerical difference between these two indices is 0.1721. This difference is, comparatively speaking, rather large, and calcite shows very strong birefringence. Crystals with strong birefringent characteristics yield a greater variety of interference colors.

OPTIC SIGN

In the discussion of birefringence, it was noted that a beam of light striking a birefringent crystal is split into two rays, the ordinary and the extraordinary ray. The direction in which the two rays travel within the crystal depends on the original direction of the beam of light striking the crystal surfaces. When the refractive index of the extraordinary ray is greater than that of the ordinary ray and, at the same time, its speed is less than that of the ordinary ray, the crystal is considered to be optically positive. If the opposite relationship holds, the crystal is considered to be optically negative.

There are several methods of determining the optic sign of a crystal. These procedures, which are based either on measuring the refractive indices of the ordinary and extraordinary rays or on obtaining the speeds of these components, are described later in this chapter.

PLEOCHROISM

A few transparent crystals exhibit an optical property called *pleochroism*. For example, if a crystal such as tourmaline is placed on the stage of a polarizing microscope, with the upper analyzer removed, it will be noticed that as the specimen is rotated on the microscope stage and viewed under plane-polarized light, a variation in color appears in the crystal specimen.   This phenomenon results from the different amount of absorption of light along the several axes of the crystal. For example, polarized light transmitted at right angles to the optic axis is more strongly absorbed, and what is generally seen is the complementary color of the original color of the crystal.   *Dichroism* is a special case of pleochroism in a crystal, involving only two colors.

EXTINCTION POSITION

When an anisotropic crystal of uniform thickness is placed on the stage of a polarizing microscope with the polarizer and analyzer crossed, it will be observed that as the stage is rotated, the crystal shows a specific color that changes in intensity as the stage is rotated. The maximum and minimum intensities of this color will be repeated four times during one complete 360° rotation of the stage.   The angle through which the crystal must be rotated to go from maximum to minimum illumination intensity (generally total darkness) is called the *extinction position* of the crystal.

For most crystals, the angle between the position of maximum illumination and the extinction position is 45°, and the angle between two consecutive extinction positions is 90°. The reason for this phenomenon of extinction and illumination is that when the crystal is in one extinction position (45°), one of the two ray components (ordinary and extraordinary) is vibrating parallel to the microscope axis, but is stopped by the crossed polarizers.   When the crystal specimen is rotated to the second extinction position, the other ray is vibrating parallel to the microscope axis, but is again stopped by the crossed polarizers.

When the crystal is rotated 45° from the extinction position, each of the ray components (ordinary and extraordinary), which are vibrating at right angles, is allowed to pass partially through the analyzer, since in this position neither ray is complete blocked.   In this position, therefore, the maximum brightness is attained.   Moreover, these two rays, which are vibrating at right angles, will vibrate in the same plane after emerging from the analyzer, although they will be traveling at different velocities.   This phenomenon can be confirmed by placing another analyzer in the crossed position above the first, a maneuver that will extinguish the light.

EXTINCTION ANGLE

The angle between a given line, edge, or plane of a crystal and the position of extinction is called the *extinction angle*. If an extinction angle is to have any meaning, the face of the crystal yielding the angle must be specified.

To measure this angle, the specimen is placed on the stage of the polarizing microscope, in a set position parallel to the crosshairs of the eyepiece and with the polarizers crossed. The stage is then rotated to the extinction position, and the angle through which it is rotated is read on the vernier. This reading, in degrees, is considered the extinction angle of the crystal.

ACCESSORIES FOR MEASURING THE EXTINCTION ANGLE

Precise determination of the extinction position of a crystal by rotation of the microscope stage, as described above, is sometimes difficult. In practice, various accessories are used in microscopy to aid in this determination and thereby to measure extinction angles exactly. Two such accessories are discussed in the following paragraphs.

Selenite Plate

A selenite plate can be used to measure the extinction angle more accurately. In use, the plate is inserted in the compensator slot of the microscope, and the crystal fragment is rotated. At the extinction position of the crystal, the only color seen through the microscope will be that of the selenite plate. If the crystal is rotated slightly from the extinction position, there will be a change in the color seen. Thus, to determine the exact extinction position, it is necessary only to rotate the stage until the color seen is exactly that of the selenite plate.

Bertrand Eyepiece

The Bertrand eyepiece (see Fig. 105) is another important accessory used to make accurate measurements of extinction angles. In use, this eyepiece requires a cap analyzer, which is placed above the eyepiece. The Bertrand eyepiece consists of four quarter-circular sections of quartz of equal thickness cut at right angles to the optic axis. Two

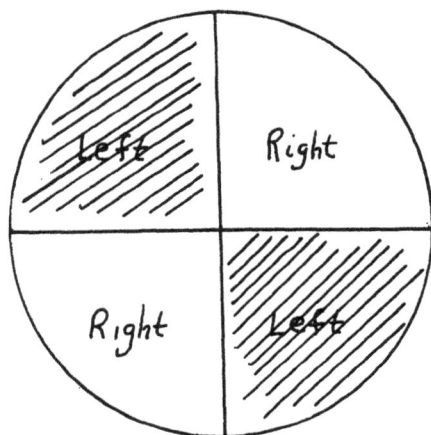

*Fig. 105.   Diagrammatic representation of the Bertrand
eyepiece (discussed in the text).*

of the sections are cut from right-hand quartz, the other two from
left-hand quartz.   The sections are mounted in the eyepiece so that
their right-angle edges are aligned with the crosshairs to make up the
four quadrants of the field of view, with left-hand and right-hand
sections alternating.

The left-hand sections rotate the plane of polarized light to the
left; the right-hand sections, to the right.   As the cap analyzer is
rotated, it reaches a position in which all four quartz sections show
a uniform light blue-green color.   In this position, the cap analyzer
is exactly at right angles to the plane of the polarized light coming
from the polarizer.

When a crystal that is not at its extinction position is viewed
through the Bertrand eyepiece, the right-hand quartz sections will
show one color, the left-hand sections another.   When the crystal
reaches its extinction position, all four quartz sections will be identical
in color.

SIGN OF ELONGATION

Many crystals display elongation along one axis (see Fig. 106).   For
example, hexagonal and tetragonal crystals are elongated parallel to

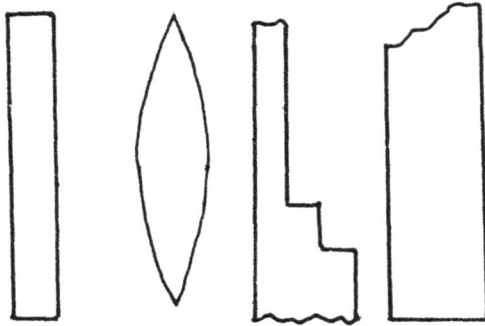

Fig. 106. Elongation
of crystals.

the *C* crystallographic axis and appear as long, regular, or broken rectangles or show prismatic shapes. The *sign of elongation* of a crystal, positive or negative, depends on the axis along which the crystal is elongated. If it is elongated parallel or nearly parallel to the fast component of the crystal (either the ordinary or the extraordinary ray, whichever has the greater velocity), it is considered to be negative in elongation and "length-fast." Conversely, if it is elongated parallel or nearly parallel to the slow component, it is considered to be positive in elongation and "length-slow."

There are some limitations to the use of the sign of elongation in identifying minerals, since some crystals may give different and inconsistent result when different sections of the crystal are examined. Nevertheless, the sign of elongation is a useful optical characteristic for describing a crystal.

INTERFERENCE FIGURES

When the Bertrand lens is inserted into the optical path of the polarizing microscope, it is converted into a conoscope with which magnified interference figures of anisotropic crystals can be observed. These interference figures consist of *isogyres* and *isochromatic curves*, which can be used in the identification and study of crystal structure. Isogyres are patterns of dark and gray areas that may or may not change as the crystal is rotated on the stage. Isochromatic curves are curved color bands or areas of color that are regularly distributed relative to the position of the isogyres. The appearance and behavior of interference figures when the stage is rotated depend largely on the orientation of the crystal.

Interference figures are valuable in the classification of crystals. They are satisfactory in identifying many anisotropic crystals, since they show characteristic appearances. By means of these figures, one can ascertain the optic sign of the crystal, the optical orientation, the degree of birefringence, whether the crystal is uniaxial or biaxial, and other characteristics.

If a section of a uniaxial crystal is cut perpendicular to its optic axis and placed on the stage of a polarizing microscope between crossed polarizers, the interference figures observed under white light consist of a dark cross and concentric colored bands (see Fig. 107, *left*). As the analyzer is turned through 90°, the dark portions become

Fig. 107. Photomicrographs of an isogyre in a uniaxial crystal (left) *and in a biaxial crystal* (right).

light and the individual colors appear to change to their complimentary colors.

Biaxial crystals show different interference figures. To prepare such a crystal for observation, a section of the crystal is cut perpendicular to an imaginary line bisecting the angle between the two optic axes. The specimen is placed on the stage of the polarizing microscope between crossed polarizers. Several characteristic interference figures are generally observed. For example, the figure may consist, with some variation, of two dark centers surrounded by concentric ribbons of color with two dark areas, in the shape of hyperbolic brushes (see

Fig. 107, *right*). As the analyzer is rotated, the colors change, as in uniaxial crystals, to the corresponding complementary colors.

Variations in interference figures, especially in isochromatic curves, depend on the thickness of the crystal, the degree of birefringence, and the optical orientation. If the crystal is very thin and allows for very little path difference of the light waves, color bands are not present, and poorly defined isogyres are seen. With a thicker crystal, the interference of the light increases, with the result that more color bands appear and the isogyres become clearer and less diffuse. Similarly, the greater the birefringence of the crystal, the more numerous the bands of color.

It must be observed that most crystal fragments are not perfectly centered with respect to their optic axis, and the interference figures, or *flash figures* as they are sometimes called, are not symmetrical. With some experience, the information gathered from observing these unsymmetrical figures can be as valuable and as informative as that obtained from perfect figures.

For example, if a uniaxial crystal is placed on the stage of a polarizing microscope (conoscopic system) with its optic axis parallel to the stage, the isogyre seen is the black cross. As the stage is rotated slightly from its original position, the cross appears to separate, forming two hyperboles (see Fig. 108). With continued rotation, the two

Fig. 108. Photomicrograph showing the separation of an isogyre in a uniaxial crystal with stage rotation.

curved segments will leave the field of view. Knowing the position of the circular stage at which, or the two quadrants of the eyepiece in which, the segments of the black cross leave the field of view will aid in determining the optic axis and optic sign of the crystal.

Color effects are also observed as the crystal is rotated. The definite variations in color and their quadrant distribution give a clue as to the position of the optic axis of uniaxial crystals.

MEASURING THE INDEX OF REFRACTION

The indices of refraction of crystals or solids are of great importance in mineralogy, chemistry, and other sciences. When the index of refraction of a transparent cyrstal or solid is to be measured with the microscope, the liquid immersion procedure is generally used. Various aids to this procedure are discussed below.

RELIEF

Viewed under the microscope, a crystal fragment immersed in a liquid of known refractive index appears irregular and perhaps pitted, and has the appearance of mountains and valleys. The appearance of these irregularities of the crystal against the surrounding medium or background is called the *relief* of the crystal.

The intensity of this relief, or its visibility against the background, depends on the refractive indices of the crystal and the surrounding liquid medium. If the refractive index of the crystal is either much less or much greater than that of the surrounding medium, the relief will be increased in intensity and show minute detail. As the two indices approach equality, the viewed relief begins to diminish. When the two indices are equal, the relief disappears. This disappearance occurs because light is neither reflected nor refracted when it travels from one medium to another of equal index of refraction, but is transmitted through crystal and surrounding medium with neither deviation nor distortion. It should be noted that in many cases when the two indices are equal, some apparent relief may be seen because of such other factors as partial absorption of the transmitted light, inhomogeneity of the fragment, or cleavage of the specimen. If monochromatic light is used, however, a transparent substance will be practically invisible when the refractive indices of the substance and the surrounding medium are the same.

BECKE LINE

The Becke line is useful in determining roughly whether the refractive index of a crystal is greater or less than the known index of the immersion liquid.

To determine the refractive index or an anisotropic fragment, the fragment is immersed in a liquid medium of known refractive index on the stage of a standard microscope. The specimen is viewed with the high-power dry objective under a narrow cone of illumination (especially of monochromatic light), which is achieved by closing down the substage iris diaphragm. The microscope is then brought to a sharp focus on the crystal fragment, whereupon a bright line concentrated around the edge of the crystal will be seen. This line is called the *Becke line*. If the microscope tube is now slightly raised or the stage slightly lowered, the Becke line moves toward the medium of higher refractive index. Thus, if the Becke line appears to move inside the crystal as the microscope tube is slightly raised or the stage slightly lowered, the refractive index of the specimen is greater than that of the immersion liquid; conversely, if the line appears to move away from the specimen into the surrounding liquid medium, the refractive index of the specimen is lower than that of the immersion liquid. If the procedure of raising the tube or lowering the stage is reversed, the Becke line effects will also be reversed.

Once the known refractive index of the immersion liquid has been determined by this method to be either lower or higher than that of the crystal, other immersion liquids with known indices successively closer to that of the specimen can be selected. Finally, by trial and error, an immersion liquid with a known refractive index exactly equal to that of the fragment under observation is found, and the fragment becomes practically invisible.

To determine the Becke line of an intact anisotropic crystal, a polarizing microscope is used. You will recall that uniaxial crystals have two indices of refraction, biaxial three. Using the polarizing microscope, it is possible to determine the refractive indices of these crystals by the Becke line method if the crystal is oriented with respect to its optic axis. This orientation can be done more easily and quickly with another accessory called the *universal stage* (see page 211).

UNIFORM CENTRAL ILLUMINATION

If crystals are fragmented or crushed into small pieces, they assume lenticular shapes and, when immersed into a liquid medium, act as convex lenses.

A crystal fragment is immersed in a liquid and then viewed under the microscope with the substage iris diaphragm somewhat closed down. The refractive index of the fragment, as compared with that of the immersion liquid, can be determined by slightly raising the microscope tube or lowering the stage. If the index of the crystal is greater than that of the liquid medium, the central part of the crystal will be illuminated; if the index of the liquid medium is greater, the central part of the crystal will be darkened.

The Becke line and uniform central illumination are generally used together to aid in determining the index of refraction of a crystal.

LIQUID STANDARDS

There are commercially available liquids with known indices of refraction that can be used as standards for determining the indices of refraction of solids. A representative listing of these liquid standards is presented in Table 3. Such standards are adequate for accurate determination of refractive indices from 1.2 to about 1.7. For determining higher refractive indices, solids with low melting points may be used.

TABLE 3. REFRACTIVE INDICES OF VARIOUS IMMERSION MEDIA

Immersion Medium	Refractive Index (at about 20°C)	Immersion Medium	Refractive Index (at about 20°C)
Air	1.000	Vinyl bromide	1.446
Methyl alcohol	1.331	Kerosene	1.448
Water	1.333	Carbon tetrachloride	1.463
Acetone	1.359	Glycerin	1.473
Ethyl alcohol	1.362	Castor oil	1.477
Heptane	1.388	Gum Arabic	1.480
Methyl butyrate	1.388	Benzene	1.501
Ethyl butyrate	1.393	Cedar wood oil	1.516
Glycerin and water (equal parts)	1.400	Anisole	1.517
Methyl cellosolve	1.403	Methyl salicylate	1.537
Amyl alcohol, 1-pentanol	1.410	Nitrobenzene	1.553
p-Dioxane	1.422	Anethole	1.562
Paraffin	1.433	1-Bromonaphthaline	1.659
Chloroform	1.446		

It should be noted that the refractive indices of liquids change with variations in temperature, generally decreasing as the temperature rises. The refractive indices of solids also change with variations in temperature, but less rapidly than those of liquids. In determining the refractive index of a solid, therefore, it is essential that the standard immersion liquid be kept at the standard temperature, usually 20°C. If the index is in doubt, it can be determined with an Abbe refractometer.

Because angles of refraction vary with varying wavelength, the refractive indices of liquids and solids show more or less variation when observed under different wavelengths of light. Liquids that show great variation when their refractive indices are determined with different wavelengths of monochromatic light are said to have *high dispersion characteristics* and are unsuitable as standards. Other liquids that show little such variation are said to have *low dispersion characteristics* and are ideal as standards. The total dispersion characteristics of a liquid can be determined by measuring its index of refraction with long-wavelength monochromatic light (such as red) and with short-wavelength monochromatic light (such as violet) and then expressing the result as the difference between the two. Most procedures specify the temperature and the wavelength of the transmitted monochromatic light (generally yellow sodium light) to be used in determining the refractive index of a substance.

Accessories for Studying Optical Properties of Crystals

QUARTZ WEDGE

The *quartz wedge* is one of the more important accessories that are routinely used in studying the properties of crystals with polarized light. The wedge, which is commercially available, consists of a piece of quartz cut in such a manner that the ordinary ray coming from it has a lower index of refraction (fast component) and is traveling or vibrating in a plane parallel to the long direction of the wedge (see Fig. 109). The extraordinary ray, having the higher index (slow component), travels at right angles to the long direction of the wedge.

The quartz wedge is used to determine the fast and slow components of crystals, the optic sign, the order of interference colors, and the degree of birefringence.

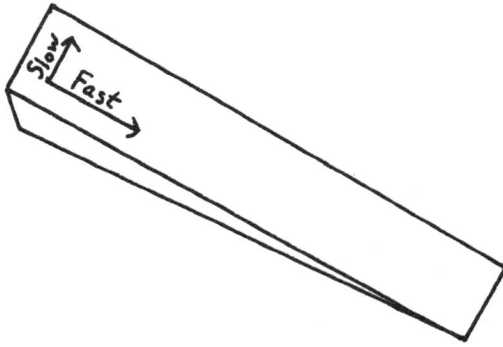

Fig. 109.   Quartz wedge.

DETERMINING ORDERS OF INTERFERENCE COLORS

The varied colors of a crystal specimen observed under the polarizing microscope indicate a few of its important characteristics and properties. Every shade and nuance of color, from vivid to pastel shades, from white to gray, is encountered in examining minerals under polarized light; the pastel shades, however, are seen most often.   In many cases, colors are used to identify mineral specimens.

A quartz wedge is inserted into the polarizing microscope in the 45° position between crossed polarizers (see discussion of extinction position, page 191), with the thin edge first.   The wedge is then moved slowly into the field along its long axis, so that a gradually thicker portion comes into view.   If white light is used for illumination, a definite change of interference colors occurs. The colors appear most intense at the thinner portion of the wedge.   The distribution of colors emerging from the wedge is caused by a path difference, which subtracts a portion of the white light.   The resulting sequences or distributions of polarization colors are referred to as the *orders of interference colors*.   The divisions into the various orders appear to come at the points where reds regularly reoccur as the thickness of the wedge increases.

The first colors seen at the thinner edge of the wedge are not sharply demarcated and are made up of a mixture of various wavelengths ranging from black through variations of gray, variations of white to pure white, variations of yellows, and finally to red. These colors are termed the *first order*.

The next range of colors observed as the rays of polarized light strike the thicker portions of the quartz wedge are more sharply demarcated, and the colors are more intense and brilliant. These pronounced colors make up the *second* and *third orders*. As the light strikes increasingly thicker portions of the wedge and the higher order of interference colors are approached, the colors grow paler and less defined until the color seen is white. The orders of interference colors are presented in Table 4.

The order of interference colors in an unknown crystal specimen is very useful in determining some of its properties, such as birefringence.

If the specimen is wedge-shaped, the order of its interference colors can be easily determined. A quartz wedge is placed at the 45° position in the compensator slot in the polarizing microscope tube. The specimen is placed on the stage, also at 45°, under crossed polarizers, with the slow direction of the quartz wedge parallel to the slow direction of the specimen.

For example, assume that an unknown wedge-shaped crystal specimen observed under these conditions shows *two* red color bands in the thin part of the crystal wedge and green as the predominant color in the thicker portions. The green color is considered a third-order interference color, since the first red color band is the demarcation point of the first order, and the second red color band marks the demarcation point of the second order (see Table 5).

Generally speaking, the higher the order of the predominant color in a crystal specimen, the greater the birefringence of the crystal, because the retardation in the wavelengths of the light becomes greater as the order of the color increases. It follows, therefore, that the color observed depends on the thickness and degree of birefringence of the crystal.

If the unknown crystal specimen is of uniform thickness, the quartz wedge is used as an accessory to determine the order of interference

TABLE 4. ORDERS OF INTERFERENCE COLORS

Order No.	Retardation (nanometers)	Interference Colors
First	0	Black
	40	Iron gray
	97	Lavender-gray
	158	Grayish-blue
	218	Pale gray
	234	Greenish-white
	259	White
	267	Yellowish-white
	275	Pale straw yellow
	281	Straw yellow
	306	Light yellow
	332	Bright yellow
	430	Brownish-yellow
	505	Reddish-orange
	536	Red
Second	551	Deep red
	565	Purple
	575	Violet
	589	Indigo
	664	Sky blue
	728	Greenish-blue

Order No.	Retardation (nanometers)	Interference (Colors)
	747	Green
	826	Lighter green
	843	Yellowish-green
	910	Pure yellow
	948	Orange
	998	Bright orange-red
	1101	Dark red-violet
Third	1128	Bluish-violet
	1151	Indigo
	1258	Greenish-blue
	1334	Sea green
	1376	Brilliant green
	1426	Greenish-yellow
	1495	Flesh
	1534	Carmine
Fourth	1621	Dull purple
	1652	Violet-gray
	1682	Grayish-blue
	1711	Sea green
	1744	Bluish-green

Order No.	Retardation (nanometers)	Interference Colors
	1811	Light green
	1927	Greenish-gray
	2007	Whitish-gray
	2048	Flesh red
Fifth	—	Greenish-blue
	—	Flesh
Sixth	—	Gray-green
	—	Pale pink
Seventh	—	Greenish-gray
	—	Pinkish-white
Higher orders	—	White

colors in the following manner: The specimen is placed on the polarizing microscope with the polarizers crossed. It is rotated until the extinction position is reached, after which it is rotated 45° from this position. Either the slower or the faster component (ordinary or extraordinary) of the specimen will now be parallel to the long direction of the quartz wedge as it is inserted in the compensator slot of the microscope, which is likewise positioned at 45°. Most quartz wedges are constructed so that the fast component (ordinary ray) vibrates in a plane parallel to the long direction.

Suppose the quartz wedge is moved very slowly, thin edge first, along its long direction over the uniform thin crystal specimen. As this is done, the order of interference colors seen will either rise or fall, depending on the relative position or orientation of the quartz wedge and the specimen. If the direction of the plane of vibration of the *fast* component of the quartz wedge is parallel to that of the *fast* component of the specimen, the order of interference colors will *rise* as the wedge is moved over the specimen, the reason being that the wedge adds thickness to the specimen and simultaneously adds to the path difference of the light waves passing through the specimen. Conversely, if the direction of the plane of vibration of the *fast* component of the quartz wedge is parallel to that of the *slow* component of the specimen, the order of interference colors in the specimen will *fall* as the wedge is moved over the specimen.

The wedge will eventually reach a position such that its fast component is exactly parallel to the slow component of the specimen, and darkness, or black, will result. At this position, the path difference produced by the specimen is exactly equal to that produced by the quartz wedge, and the wedge is said to *compensate* the interference color of the crystal. The position of compensation can therefore aid in determining the order of interference colors of the specimen. If the position of compensation cannot be obtained at first—that is, if the order of interference colors rises as the wedge is moved across the specimen—rotate the stage 90° and try again.

Once the position of compensation has been found, the order of interference colors can be determined by withdrawing the wedge slowly, noting the sequence of colors, and comparing them with the orders of interference colors listed in Table 4. For example, a crystal specimen of uniform thickness is placed on the stage of the polarizing microscope at 45° from the extinction position. Suppose the predominant color of the specimen is orange. The order to which this interference color belongs must now be determined. To make this determination, the quartz wedge is moved slowly, thin edge first, along its long direction over the specimen. Suppose that the observed interference colors go down the scale in the following order: orange, green, blue, indigo, purple, red, yellow, grayish-white, and black (the point of compensation). The orange color must therefore belong to the second order (see Table 4). If the crystal specimen is removed and the quartz wedge slowly withdrawn, the colors observed in the wedge alone will be the same as those just listed.

DETERMINING THE DEGREE OF BIREFRINGENCE

The birefringence of a crystal can be roughly calculated from the equation:

$$\text{Birefringence} = \frac{\text{retardation (order of dominant interference color)}}{\text{thickness}}$$

from which it can be seen that a crystal with a higher order of interference color will show higher birefringence, or conversely, that a crystal with high birefringence will yield interference colors of higher orders on the scale.

For example, assume that one wishes to determine the degree of birefringence of a certain crystal specimen. The thickness of the specimen is 0.030 mm, or 30,000 nanometers, and the interference color is green of the second order. From Table 5, the retardation, or path difference, of this color is 747 nanometers. The birefringence of the crystal can be calculated as follows: birefringence = 747/30,000 = 0.025. The value 0.025 is an absolute quantity. It represents a medium degree of birefringence for the crystal in question, as compared to the value of 0.172 for a calcite crystal, which is highly birefringent.

DETERMINING THE OPTIC SIGN

The quartz wedge is used not only to determine the order of the dominant interference color of a crystal and its fast and slow directions, but also, in conjunction with interference figures, to determine the optic sign of uniaxial crystals.

The specimen is placed on the stage of the polarizing microscope, and the Bertrand lens is inserted in the light path. As the stage is rotated slowly, the isogyre or black cross, which is generally seen, splits into two hyperboles and then slowly leaves the field of microscope *along* the C crystallographic axis. The C axis is an imaginary line or direction parallel to the optic axis of the crystal; its position can be easily located with the crosshairs in the eyepiece. These crosshairs divide the field of view into four quadrants, by means of which the C axis can be pinpointed. The extraordinary ray or component vibrates parallel to the C axis, the ordinary component perpendicular to it.

After identifying and noting the quadrants that locate the C axis, remove the Bertrand lens from the optical path, rotate the specimen 45° from this position, and cross the polarizers. Insert the quartz wedge in the 45° position. Gradually move the wedge over the crystal specimen and note whether the fast direction of the wedge is vibrating parallel to the C axis. If the order of interference colors increases, the fast direction of the quartz wedge is vibrating parallel to the fast-direction component (extraordinary ray) of the crystal, which is therefore considered optically negative. (You will recall that, by definition, if the speed of the extraordinary ray in a crystal is greater than that of the ordinary ray, the crystal is said to be optically negative.) Conversely, if the order of interference colors goes down the scale, the speed of the ordinary ray is greater, and the crystal is optically positive.

SELENITE (GYPSUM) PLATE

The *gypsum* or *selenite plate* (see Fig. 110) is another accessory used with the polarizing microscope. It is made of selenite, a transparent crystalline variety of clear gypsum, cut or split to a specific thickness to give a first-order red interference color. (Actually, the color looks red-violet with a retardation value of one wavelength.) The plate is mounted in a metal holder and oriented so that the plane of vibration of its fast component is parallel, and that of its slow component is at right angles, to the long direction of the holder. Arrows are inscribed on the metal holders of most commercial selenite plates to indicate the fast and slow directions.

The selenite plate is used in the same manner as described for the

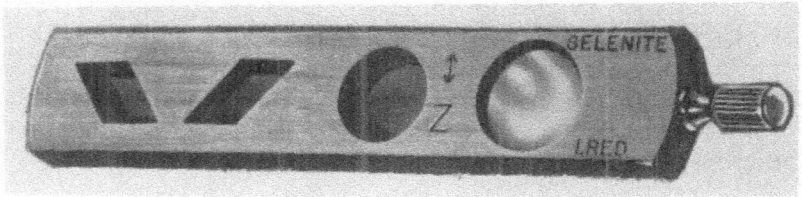

Fig. 110.   Selenite (gypsum) plate.

quartz wedge to determine the slow and fast directions and the optic sign of a crystal.   It is very sensitive to any slight change in retardation or interference and is therefore used for examining thin crystals or crystals that have low birefringence and low retardation.

MICA PLATE

The *mica plate* (see Fig. 111), also called the *quarter-wavelength retardation plate,* is another accessory that can be used instead of the

Fig. 111.   Mica (quarter-wavelength retardation) plate.

selenite plate, particularly in examining crystals that have high retardation and that give second- to fourth-order interference colors. Because the retardation of the mica plate is only a quarter wavelength, the colors observed in such crystals move only a short distance on the interference color scale when this type of plate is used.   On the other hand, use of a selenite plate, with its retardation of about half a wavelength, would cause an interference color change of a whole order, and a change of this magnitude would make determination of the exact order of interference colors of crystals with high retardation effects somewhat difficult.

The mica plate is a sheet of mica of appropriate thickness to produce a path difference of a quarter wavelength under a sodium yellow light. Its interference color, seen with white light, is light gray. Mounted in a metal holder, it is generally oriented so that the direction of its fast component is parallel to the long direction of the holder. The mica plate holder is usually marked with arrows to show the fast and slow directions.

The mica plate is used in a manner similar to that described for the quartz wedge and selenite plate to determine the directions of the fast and slow components of crystals, as well as to determine the optic signs of crystals from interference figures.   With the mica plate, black dots appear in place of the black curves of the interference figures.   In general, a mica plate or a quartz wedge should be used for crystals having interference colors above first-order yellow, a selenite plate for crystals having first-order colors such as white or gray.

BEREK COMPENSATOR—MEASURING THE PATH DIFFERENCE

The *Berek compensator* (see Fig. 112) is used to measure accurately the path or phase difference of a birefringent crystal specimen.   This instrument consists of a calcite plate about 0.1 mm thick, cut at right angles to its optic axis and mounted in a metal holder that acts as the shaft.   At one end of this shaft is a calibrated drum, graduated in nanometers, that measures the angular rotation of the calcite plate.   A built-in magnifier facilitates reading this rotation on the drum scale.

In use, the Berek compensator is first placed in the accessory slot in the microscope body tube.   The crystal specimen is rotated on the microscope stage to determine the direction of the fast component, which is then positioned parallel to the slow direction of the calcite plate.   (The slow and fast directions are inscribed on the compensa-

Fig. 112. Berek compensator.

tor.)   The calcite plate is now moved until compensation is achieved, and the path or phase difference in nanometers is read on the drum. A monochromatic source of illumination (sodium D-line, wavelength 589 nanometers) is preferred for more accurate measurements of path difference.   An interference filter that transmits light with a wavelength of 589 nanometers is commercially available.

The Berek compensator is also used to determine the fast and slow components and the optic sign of a crystal in a manner similar to that described for the selenite plate.

UNIVERSAL STAGE

The *universal stage* (see Fig. 113) permits rapid orientation of a crystal so that accurate determinations of crystal properties can be made.   This instrument, which can be rotated around four or five axes, allows the mineral specimen to be oriented in various angular positions relative to the plane of the microscope stage.   It is especially useful in evaluating the optical properties of complicated biaxial crystals because it allows ready determination of the optic orientation

Fig. 113.   Universal stage.

of the crystal, which is especially important in measuring the two refractive indices of a biaxial crystal.

In using the universal stage, the general procedure is to place one or
two drops of an immersion fluid (glycerin) in the center of a circular
glass plate that is part of the stage. The crystal specimen slide is placed
on this glass plate with the specimen directly over the immersion fluid.
A cover glass is placed over the specimen and an additional one or
two drops of the immersion medium is placed on top of the cover
glass.

The detachable upper part of the universal stage, which is an optical
hemisphere, is placed on the slide over the specimen and then clamped
to the stage by two thumbscrews.   A few drops of immersion fluid is
also placed on the lower part of the glass plate, and the lower optical
segment is then attached to the underside of this plate.   The function
of these optical segments is to eliminate any refraction of the light
rays entering and leaving the crystal specimen.   These segments are
purchased in pairs, each pair having the same index of refraction. The
refractive index of the pair used to examine a given crystal specimen

should be approximately equal to that of the specimen. The immersion fluid eliminates total reflection of the light. With the universal stage, it is a simple matter to rotate the crystal specimen so that it lies in the desired position.

EXAMINING CRYSTAL SPECIMENS WITH THE POLARIZING MICROSCOPE

Preparing Crystal Specimens

The polarizing microscope aids in the identification and determination of the optical and physical properties of crystals and nonopaque objects. Crystal specimens must be prepared in advance. Generally, crystals or fragments are reduced in size by grinding with a pestle in an agate or steel mortar (see Fig. 114). Certain crystals, however, such as those showing cleavage, may require special treatment. A

Fig. 114. Mortar and pestle.

crystal with cleavage is placed in a mortar and struck with sharp blows until small crystal specimens with flat faces are obtained. Crystals with fibrous properties are held between a vise and then cut into thin, transparent layers with a sharp blade. Thin sections of ore minerals can be cut from a large section and then ground and polished, as described in Chapter 10. Very thin sections can be mounted in balsam on glass slides.

In most cases, however, the specimen is reduced to a fine powder by grinding in a mortar. If specimens of a uniform size are desired, the resultant powder may be passed through a 60-mesh screen onto a 100-mesh screen. If a finer specimen is desired, the powder is screened through a 100-mesh screen onto a 200-mesh screen. (The expressions 60-, 100-, and 200-mesh mean that the screens have, respectively, 60, 100, and 200 holes to the inch.) The ground crystalline material may also be examined microscopically without sieving.

A few drops of a standard liquid of a known refractive index is placed in the center of a glass slide, and a small amount of the crystalline material is then mounted in the immersion medium. This is done by placing the powder at the end of a spatula and tapping the spatula until sufficient material is added. Care must be taken not to add too much of the material, since the crystals may clump together and be difficult to examine. A cover glass is then placed over the crystalline material to protect it.

A single specimen grain, located under the microscope, can be isolated and manipulated with a needle so that it is oriented in the correct position. The crystal can also be transferred from slide to slide containing different immersion media when the refractive index of the crystal is to be determined.

Adjusting the Microscope

The polarizing microscope must be in perfect working order before it can be satisfactorily used. It must be kept at room temperature to prevent strain in the optical elements of the instrument, which may produce distortion. The microscope should be examined regularly and slight adjustments made as required.

CENTERING THE OPTICAL PARTS

The eyepiece and objectives must be exactly centered in the polarizing microscope, since measurements of the optical properties of crystals will be inaccurate if the optical system itself is out of alignment. To center the system, first center the low-power dry objective. Then place a stage or disc micrometer on the stage of the microscope and adjust the micrometer on the stage so that the intersection of the crosshairs is directly over the zero marking. Rotate the stage slowly and note whether the intersection of the crosshairs remains over the zero. If it does not, adjust the centering screws attached to the center-ing ring of the objective until it does. This method is more fully illustrated by the following example: Assume that as the stage is rotated slowly, the zero appears to move outward and upward until it reaches a maximum position and then, with continued rotation of the stage, reverses itself and starts to move downward toward the inter-section and beyond the vertical line of the crosshairs. When this position is reached, stop the stage and, with the centering screws, move the zero image vertically about half the distance toward the intersection of the crosshairs. (Do not move the image all the way down to the intersection.) Now move the stage micrometer manually until the original zero is exactly positioned at the intersection of the crosshairs. It will be found that as the stage is rotated, the zero will remain at the intersection of the crosshairs and appear to be rotating around the midpoint. Do not touch the centering screws after the objective is centered.

Centering of all the other available objectives is done in the same way, but without moving the stage or disc micrometer, so that the zero will be centered at the same point for all the objectives. Center-ing of these remaining objectives is done by simply manipulating the centering screws of each objective.

With the micrometer on the stage and the low-power objective cen-tered, check the eyepiece for centering by rotating it. Note whether the intersection of the crosshairs remains approximately within the zero. An eyepiece cannot be adjusted; if it is not centered properly, it should be returned to the supplier for repair.

The analyzer should be aligned in the polarizing microscope in the proper relationship to the polarizer. Check the proper setting in the following way: Remove the upper condenser from the optical path. Remove both the objective and eyepiece and insert the analyzer in the optical path. Set the polarizer at zero and then rotate the analyzer until the field appears in complete darkness. The setting on the analyzer should be at zero. Any slight rotation of the analyzer from this position should cause the field to become brighter. If the setting is not at zero, adjust the analyzer so that the setting is zero at the maximum darkness of the field. At this position, the polarizer and analyzer must be at the zero setting.

The Bertrand lens must be accurately centered for use.   Place a quartz crystal fragment, mounted on a glass slide, on the stage of the microscope and bring the high-power dry objective close to the slide, almost touching it.   Set the analyzer and polarizer at their zero positions.   In this position, the darkness of the field will be at its maximum.   Open the iris diaphragms of both condensers (substage and auxiliary), and place the Bertrand lens in the optical path of the microscope.   Focus the Bertrand lens so that the interference figure in the back lens of the objective is sharp.   The interference figure seen is an isogyre (concentric rings of color surrounding a large black cross), the center of which should exactly lie at the intersection of the cross-hairs in the eyepiece.   If it does not, adjust the position of the inter-ference figure by manipulating the centering screws attached to the Bertrand lens.

If further adjustment is needed before the microscope will function properly, it would be best to send the instrument back to the factory for the proper settings.

Studying Optical Properties of Crystals

After the polarizing microscope has been properly adjusted, the various properties of a crystal specimen can be determined by the procedures outlined below.

A. With the analyzer swung out of the optical path, and the crystal specimen immersed in a standard liquid of known refractive index, perform the following:

1.   Determine the refractive index of the crystal relative to the refractive index of the standard immersion liquid.

2.   Examine the color of the crystal and determine whether the specimen shows pleochroism under transmitted light.

3.   Observe the physical appearance of the crystal, that is, the crystallization shape, the cleavage, if any, and other noteworthy characteristics.

B.   Insert the analyzer in the optical path in the crossed position and perform the following:

1. Rotate the specimen on the microscope stage and determine whether the crystal is isotropic or anisotropic.

2. If the crystal is isotropic, determine its single refractive index, using white light or monochromatic light of a specified wavelength and standard immersion media of known refractive indices.

3. If the crystal is anisotropic, observe the interference colors and determine the degree of birefringence in relationship to the thickness and orientation of the crystal.

4. With the Bertrand lens in position, examine the interference figures of the anisotropic crystal to determine whether it is uniaxial or biaxial.

5. Determine the appropriate optical properties of the anisotropic crystal, such as optic sign (see Fig. 115), extinction angle, refractive index, degree of birefringence, and sign of elongation.

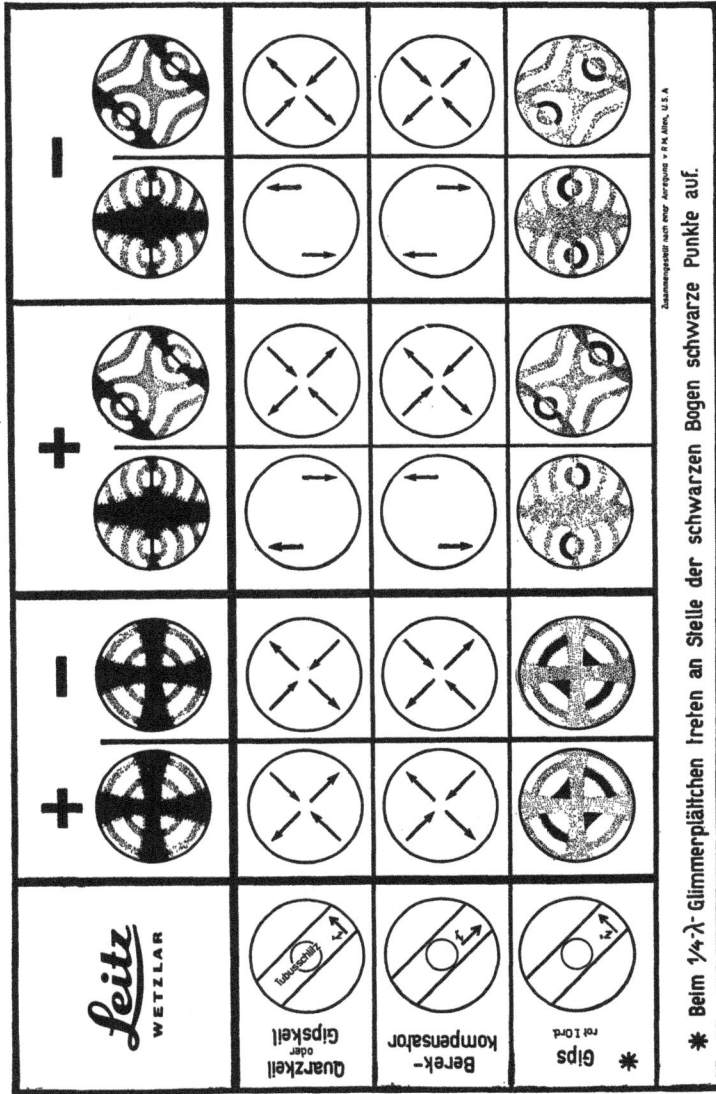

Fig 115. *Chart: Determination of optic sign by means of the conoscope interference figures appearing with various types of compensators (arrows show the directions of displacement of the figures with increasing path difference in the compensator).*

Quartz or gypsum wedge ▼

Berek compensator ▼

Gypsum red 1st order ▼

With the ¼ W. L. mica plate black dots appear in place of the black curves. ▼

* Beim ¼-λ- Glimmerplättchen treten an Stelle der schwarzen Bogen schwarze Punkte auf.

CHAPTER 8

The Phase-Contrast Microscope

THE PHASE-CONTRAST MICROSCOPE is a recent valuable development in microscopical science. Its greatest value has been in examining transparent or nearly transparent live tissues or structures that cannot ordinarily be seen with the standard compound microscope under bright-field illumination. For example, mitochondria and chromosomes —transparent structures that are difficult to stain and are ordinarily invisible in most microscope cell slides—can be seen plainly with the phase-contrast microscope.

As mentioned in Chapter 7, an understanding of a few fundamental concepts of the nature and behavior of light is essential to efficient use of the phase-contrast microscope. To this end, you might find it advantageous to review the discussion on the nature of light in that chapter before reading further.

GENERAL PRINCIPLES

If the index of refraction of a specimen differs only slightly from that of the surrounding medium, the specimen will be invisible or nearly so because it will not show relief, unevenness, or contrasting difference. Essentially, phase-contrast microscopy is a special method of providing controlled illumination to improve the visibility of such specimens. This improvement is effected by increasing the index of refraction either of the specimen itself, so that it appears dark against a lighter background, or of the surrounding medium, so that the specimen appears light against a darker background. Such controlled illumination makes it possible to observe the interferences in the optical path of the light caused by the varying thickness of the specimen; these interferences increase contrast within the specimen, so that more details can be readily seen. There are many devices used in phase-contrast microscopy to obtain different views of the same image of the specimen in order to interpret its structure correctly. Basically, all these devices take advantage of any small retardation of light waves traveling through a specimen that has an index of refraction either slightly higher or slightly lower than that of the surrounding medium

By way of illustration, consider the problem of a specimen with an index of refraction slightly higher than that of the surrounding medium. Light waves traversing any particle of such a specimen are somewhat refracted, causing them to be deviated and slightly retarded in speed as compared to those traveling through the surrounding medium. The result is that light waves coming from the particles of the specimen are out of phase with those coming from the medium (see Fig. 116). It has been found that if the difference between the index of refraction of the specimen particle and that of the surrounding medium is slight, the retardation of the light will also be slight, producing a path difference of about a quarter wavelength. Consequently, there will be two waves entering the objective of the microscope after passing through the specimen and the surrounding medium—one retarded by the specimen, the other unretarded by the medium. Under these circumstances, since the eye is not sensitive to a change of phase in light waves, and since there is no contrast between a transparent specimen and the surrounding medium, such a specimen would be invisible with the standard compound microscope. It is true that blocking of light rays from the surrounding medium, as is done in dark-field illumination, would provide additional contrast. Dark-field illumination, however, enhances surface details, but not internal details. Total blocking of light rays from the specimen, on the other hand, would so darken it that no details could be seen.

This problem of visibility of a transparent or nearly transparent specimen was solved when it was found that the contrast of the image could be intensified by changing the retarded light waves transmitted through the specimen. The change made is in the phase of the light waves, and it is such as to increase the amplitude of the light waves reaching the eye. Although the eye is not sensitive to a phase difference

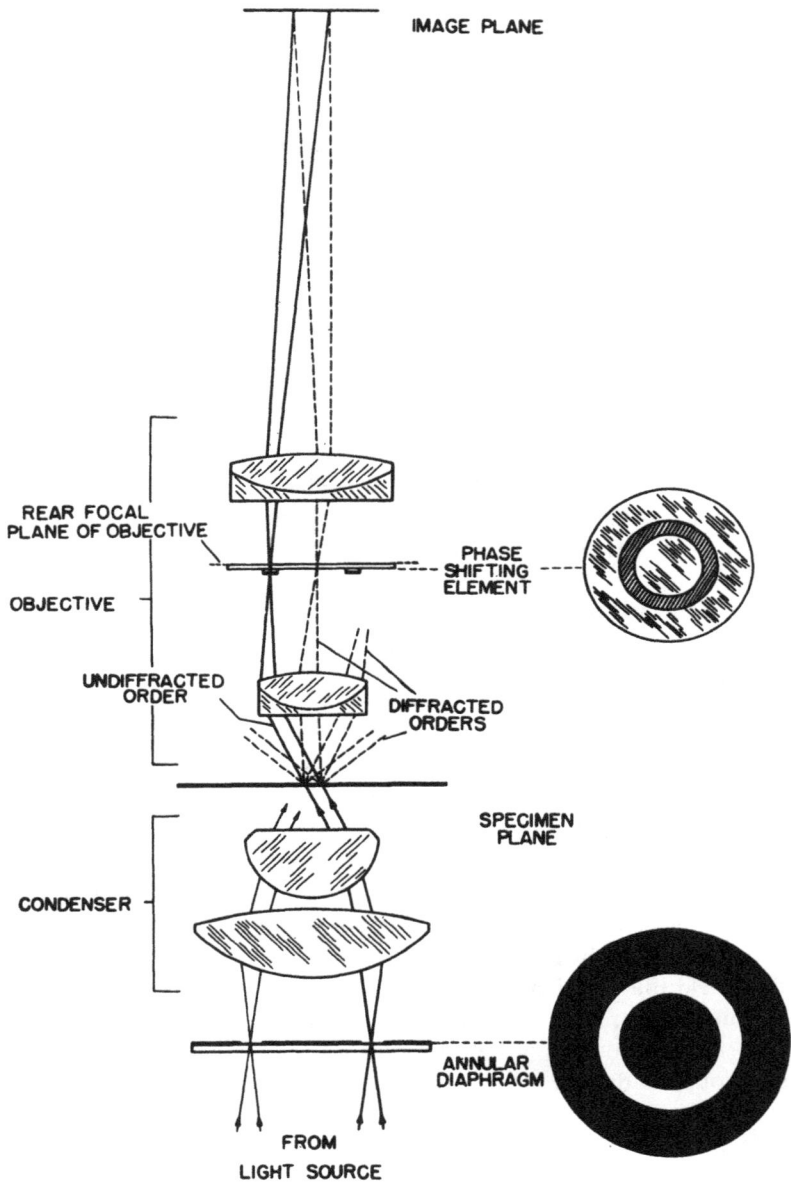

IMAGE PLANE

REAR FOCAL
PLANE OF OBJECTIVE

PHASE
SHIFTING
ELEMENT

OBJECTIVE

UNDIFFRACTED
ORDER

DIFFRACTED
ORDERS

SPECIMEN
PLANE

CONDENSER

ANNULAR
DIAPHRAGM

FROM
LIGHT SOURCE

Fig. 116. Diagrammatic representation of the path of light through a phase-contrast microscope.

in light waves, it is sensitive to a difference in amplitude (brightness). Thus, the increase in amplitude causes an increase in the brightness, and therefore, the contrast, of the specimen.

Now, recall the earlier statement that light waves transmitted through a specimen with an index of refraction either slightly higher or slightly lower than that of the surrounding medium will be out of phase by about a quarter wavelength with those transmitted through the medium. Most microscopic living organisms fulfill these requirements. If the retarded light waves coming from such a specimen are advanced, or if the unretarded light waves coming from the medium are retarded a quarter wavelength by artificial means, the specimen will appear brighter than its background, thus producing a great increase in contrast.   Conversely, if the retarded light coming from the specimen is further retarded, or if the unretarded light coming from the surrounding medium is advanced a quarter wavelength by artificial means, the specimen will appear darker against a lighter background because of some annulment of the light waves, thus producing a similarly great, though reversed, increase in contrast.

The aim of phase-contrast microscopy, then, is to alter the path difference of the light waves from the specimen and those from the surrounding medium so as to change the amplitude of the light waves and thereby produce visible changes in the image of the specimen.

CHANGING PHASE DIFFERENCE ARTIFICIALLY

Diffraction Grating (Phase Plate)

A diffraction grating, or phase plate (see Fig. 117), incorporated within the special phase objective used in the phase-contrast microscope serves to artificially retard light waves transmitted through the objective. The grating is a piece of optical glass on which thousands of equally spaced parallel groves are ruled. It functions by diffracting light rays in such a way as to cause the desired interference.   This diffraction can be thought of as being similar to that occurring when light strikes the particles within a specimen and the surrounding medium.   (To put this another way, the grooves of the grating can be regarded as the particles of a specimen and the elevations between

LEGEND

 Glass

 Phase retarding material

 Absorbing material

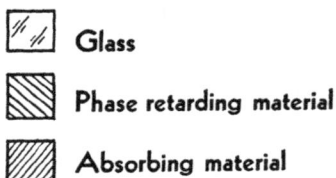

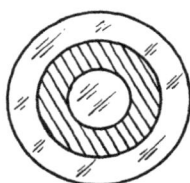

Diffraction Plate

Bright Contrast   Dark Contrast

Fig. 117. Diffraction grating (phase plate): top view (A), showing conjugate and complementary areas; cross-sectional views, showing areas coated for bright contrast (B) and dark contrast (C).

the grooves as the surrounding medium, or vice versa.) In this way, the grating effects the complete or partial separation of the deviated and undeviated light waves coming from, respectively, the specimen and the surrounding medium—thereby enhancing the amplitude (brightness) of either the deviated or the undeviated waves—and refocuses them on the image plane for the eye to see.

The desired constructive interference of the light waves occurs in the following way: Light emerging from the microscope condenser, of course, strikes both specimen and surrounding medium. The light coming from the specimen is diffracted and deviated because the index of refraction is slightly greater than that of the medium. The light emerging from the surrounding medium, on the other hand, is transmitted with almost no deviation. The deviated light waves from the specimen are brought to a focus at the focal plane of the eyepiece by the microscope objective. The undeviated light waves from the medium are brought to a focus at the rear focal plane of the microscope objective. A diffraction grating is inserted at this position, in the rear focal plane of the objective, to change the phase or optical path of either the deviated or the undeviated rays and thus separate them.

The grating, in addition to being ruled, is coated with a thin film of a refractive material such as magnesium fluoride, which is deposited by means of a high vacuum. The thickness of this deposited film is calculated precisely to produce a phase shift of a quarter wavelength of green light, this thickness usually being sufficient to ensure retardation of the undeviated light waves from the surrounding medium by a quarter wavelength, so that both constructive interference and greater contrast result.

There remains, however, the problem that the deviated light from the specimen is weaker than the light from the surrounding medium, a situation that would result in decreased contrast. To remedy this, a very thin coating of the alloy Inconel (78% nickel, 15% chromium, 6% iron) is superimposed on the magnesium fluoride film. The Inconel absorbs a portion of the light from the medium, so that the intensities of the deviated and undeviated light are equalized and maximum contrast is achieved.

It may be necessary to vary the contrast for living materials that require weaker or stronger contrast for best resolution. The variation can be obtained by using gratings with different absorption characteristics.

CONJUGATE AREA (PHASE-ALTERING ANNULUS) AND COMPLEMENTARY AREA

When a diffraction grating is optically centered and positioned in the rear focal plane of the objective and the substage iris diaphragm is closed down to a small aperture, the image of the aperture will be focused in a small area on the grating. This area, a ring-like band that coincides with the image of the iris diaphragm aperture, is called the *conjugate* or *phase-altering annulus*. The remaining area is called the *complementary area*. The undeviated light waves are transmitted through the conjugate area, the deviated waves through the complementary area.

Since the shape and dimensions of the iris diaphragm aperture determine the shape and dimensions of the conjugate area and the formation of interference effects, which determine the nature of the final image, a special device called an *annular aperture diaphragm* (see Fig. 118) is placed at the focal plane of the substage condenser. The

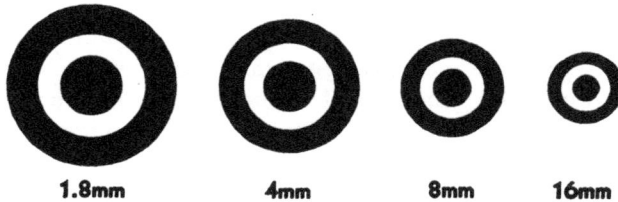

1.8mm   4mm   8mm   16mm

Fig. 118.   Annular aperture diaphragms.

light from this special diaphragm illuminates the specimen with a limited cone of light and also controls the light passing through the surrounding medium.

Bright and Dark Contrast

If the aforementioned refractive-material coating of precise thickness is deposited on the conjugate area, the undeviated rays passing through this area are retarded by a quarter wavelength. This retardation produces constructive interference in the deviated waves from the specimen and the undeviated waves from the surrounding medium, so that the resultant waves, which delineate the specimen image, are of greater amplitude than either the undeviated or the deviated waves, and the specimen appears brighter against a darker background. The result is called *bright* or *negative contrast.*

If the phase-retarding material is deposited on the complementary area, the deviated waves passing through this area are retarded a half wavelength (one quarter original deviation plus an additional quarter). This retardation produces destructive interference in the deviated and undeviated waves, so that the resultant waves, which delineate the specimen image, are of lesser amplitude than either the deviated or the undeviated waves, and the specimen appears darker against a lighter background. The result is called *dark* or *positive contrast.*

Although both methods improve contrast, it has been found that bright contrast gives better resolution and definition of the specimen.

Dark-contrast objectives nevertheless remain popular, since the results achieved are similar to those achieved with stained specimens.

It should be noted that if there is more than slight variation in the optical differences of different specimens, an assortment of gratings may be required to achieve the desired contrast for each specimen. In most cases, however, a single grating will suffice for specimens that show a small range of optical differences.

Furthermore, the resolving power of the microscope is impaired slightly by a diffraction grating, and the grating may produce halos around the specimen. The substantial increase in contrast achieved with the phase-contrast microscope, however, more than compensates for any slight loss of resolution.

PHASE-CONTRAST MICROSCOPY

The effectiveness of phase-contrast microscopy depends essentially on how fine a separation of the deviated from the undeviated light waves can be achieved. For optimum results, the opening of the annular aperture diaphragm must be circular and exactly centered on the optical axis of the microscope. Under these conditions, the diaphragm will transmit all the undeviated waves and only a small portion of the deviated waves through the conjugate area of the grating; the major portion of the deviated waves will pass through the complementary area.

If necessary, a standard compound microscope can be converted to a phase-contrast microscope by inserting a diffraction grating in the focal plane of the objective and an annular aperture diaphragm below the substage condenser. The essentials of phase contrast described above are, however, achieved with the phase-contrast microscope, which is expressly designed for the purpose.

Most standard phase-contrast microscopes are similar in design and arrangement of the essential parts. In such a microscope (see Fig. 119), an annular aperture diaphragm, with centering screws, is mounted below the substage condenser, at its focal plane, so that the specimen is illuminated with a hollow cone of light. With each phase objective

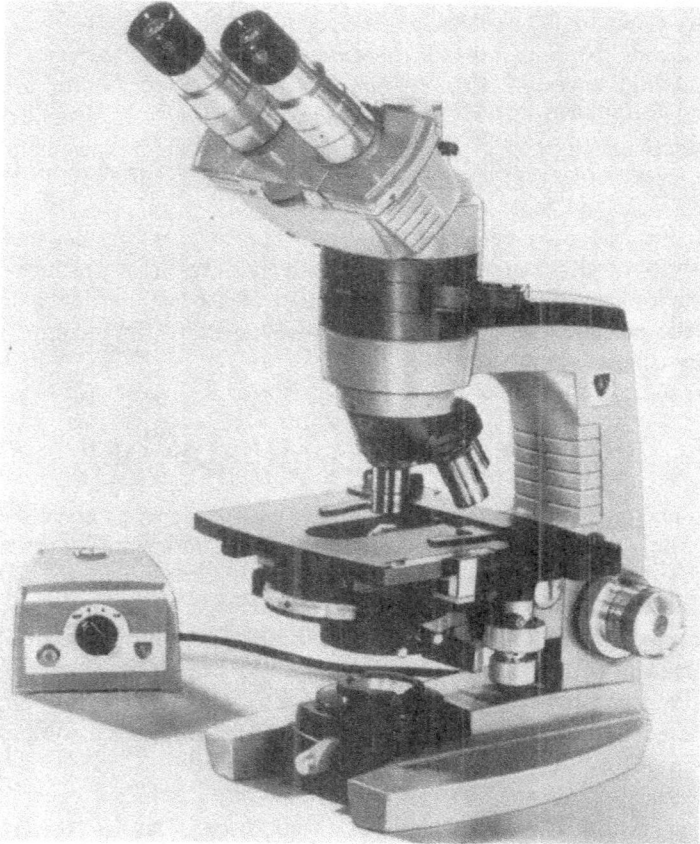

Fig. 119. Phase-contrast microscope.

of different magnification, a different diaphragm is used for optimum results. A rotating turret plate carrying three or four such diaphragms can be quickly mounted below the condenser at its focal plane; the turret need only be rotated to bring the annular diaphragm with the desired aperture into the optical path. Such a turret plate, used with a revolving nosepiece carrying several different phase objectives, facilitates rapid changing of phase objectives and corresponding annular diaphragms.

The diffraction grating is usually designed as an integral part of the objective. The incorporation of gratings in the phase objectives, incidentally, does not necessarily preclude the use of these objectives in normal microscopy. Phase objectives are available in wide variety; the selection includes objectives designed to provide bright or dark contrast, each with three degrees of contrast—high, medium, and low. They can be screwed into a revolving nosepiece of the type mentioned above.

The phase-contrast microscope is equipped with a separate built-in adjustable telescope, which consists of the eyepiece combined with a Bertrand lens.   This telescope is used to bring into focus the coating of the grating positioned at the rear focal plane of the objective and to ascertain whether the image of the diaphragm annulus is in coincidence with the conjugate area (phase-altering annulus).   The Bertrand lens can be raised or lowered for focusing and removed from the optical path when it is not in use.   The use of this lens is similar to its use in the polarizing microscope.

The other mechanical and optical parts of the phase-contrast microscope are similar to those of the standard compound microscope.

Illumination for the Phase-Contrast Microscope

Since magnesium fluoride is more sensitive and has a greater index of refraction in the blue region of the spectrum than in the red, diffraction gratings, with their deposited films of magnesium fluoride and Inconel, are not achromatic for all wavelengths of light.   Thus, if white light is used with these gratings, there may be divergences from the desired quarter-wavelength deviation because of blue and red wavelengths. Most gratings are therefore standardized for use with green monochromatic light.   Ideally, monochromatic rather than white light should be used, so as to decrease the amount of deviated waves impinging on the conjugate area. If white light is to be used, a green filter is placed in front of the light source to improve contrast.   It is essential to use the Köhler method of illlumination with the phase-contrast microscope for optimum results.

It is also possible to use two contrasting colors with the annular aperture diaphragm, so that undeviated light of one color passes through the conjugate area of the grating and deviated light of the second color through the complementary area. Under these circumstances, a colored image of the specimen is formed. Bright contrast gives a brightly colored specimen surrounded by a dark halo of a different color; dark contrast, a darker colored specimen surrounded by a brighter halo.

There are various methods of achieving color phase contrast. One is to replace the standard annular diaphragm with a bicolored one. A filter of one color replaces the dense part of the diaphragm; a filter of contrasting color replaces the band on the diaphragm that transmits the light. The bicolored annular diaphragm is used in conjunction with a standard neutral diffraction grating.

As with the standard compound microscope, the resolution of a specimen under the phase-contrast microscope will be increased if shorter wavelengths of light are used. Ultraviolet light using quartz objectives will increase resolution and contrast substantially. In many cases, details of the specimen are made visible on film, whereas no contrast appears with ordinary light.

Adjustment of the Phase-Contrast Microscope

The optical system of the phase-contrast microscope must be aligned properly with the source of light if the instrument is to perform efficiently. The light source, which must be sufficient to illuminate the specimen uniformly, must be centered in relation to the microscope mirror, or, if a mirror is not used, the light must be in alignment with the optical axis of the condenser, objective, and eyepiece.

It is essential that the light coming from the transmitting portion of the annular aperture diaphragm, that is, the aperture image, be in coincidence with the conjugate area of the diffraction grating. It must be emphasized that the effectiveness of any phase contrast microscope depends on the degree to which this coincidence is achieved. Furthermore, the resolution of the specimen and the contrasting effects achieved depend greatly on correct microscope alignment, with the conjugate and complementary areas of the grating sharply demarcated.

General Procedure for Using the Phase-Contrast Microscope

The general procedure for using the phase-contrast microscope is as follows:

Step 1. Assemble the required annular diaphragms, phase objectives, and eyepieces, and position them in the microscope.

Step 2. Place the specimen slide on the microscope stage and adjust the instrumemt for Köhler illumination (see page 42). It is important that the cover glass be of the correct thickness, since too great a thickness may cause spherical aberration within the objective.

Step 3. Position the illuminator so that the iris diaphragm of the illuminator (field diaphragm) is from 6 to 9 in. from the substage mirror. Align the lamp and adjust the mirror so that the light from the lamp is centered in the mirror and reflected upward through the optical axis of the microscope.

Step 4. Using the rack and pinion, position the substage condenser so that the light passing through the condenser and the objective falls entirely within the conjugate area of the diffraction grating. Open the substage iris diaphragm so that its aperture is slightly larger than that of the annular aperture diaphragm. If the body tube is adjustable, adjust it to the exact length required for the objective in use.

Step 5. Remove the eyepiece and bring the auxiliary telescope device (Bertrand lens) into the optical path of the microscope; then bring the coating of the diffraction grating into focus. Adjust the annular aperture diaphragm by means of the centering screws so that the image of the ring-like opening is positioned exactly within the conjugate area of the grating.

Usually, the phase-contrast microscope is first aligned using the blank part of the slide, since the specimen itself may cause excessive deviation of the light in the complementary area.

Step 6. Remove the centering telescope from the optical path, replace with the eyepiece, and bring the specimen into sharp focus. If the center of the field appears dark and the field is unevenly illuminated, adjust the substage condenser by racking it up or down until the field is evenly illuminated.

Step 7. Close down the substage iris diaphragm and focus the image of the lamp filament as sharply as possible on the iris diaphragm or on the annular aperture diaphragm positioned below the substage condenser. This is accomplished by manipulating the variable focusing condenser attached to the illuminator. If the lamp condenser is not focusable, move the illuminator itself to achieve focus. If the illuminator housing is supplied with a groundglass plate, remove this plate temporarily while adjusting the lamp. Adjust the mirror of the microscope during focusing so that the filament bulb image is exactly centered in the mirror. The image of the lamp filament should fill the opening of the annular aperture diaphragm.

Step 8. Open the substage iris diaphragm so as to allow sufficient light to illuminate the specimen for good observation. Remove the eyepiece, replace with the centering telescope, and again align the optical axis of the annular aperture diaphragm with the optical axis of the objective so that the annular image is exactly centered and positioned within the conjugate area (see Fig. 120). This sharp adjustment can be observed when there is a sharp delineation between the conjugate area and the complementary area, with very little overlap.

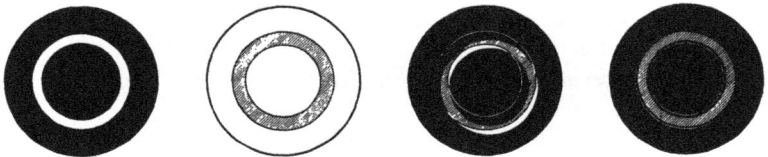

Fig. 120. Aligning the annular aperture diaphragm and the diffraction grating: (left to right) *diaphragm, poor alignment, and good alignment.*

Step 9. Replace the centering telescope with the eyepiece. Close or open the lamp diaphragm (field diaphragm) until the area of illumination of the lamp image is about equal to or slightly less than the area of the microscope iris diaphragm. Adjust the mirror, if necessary, so that the lamp image is centered in the viewed field of the microscope.

The adjustment of the field diaphragm for the phase microscope differs at this point from the adjustment for the standard compound microscope. If the field diaphragm is closed down too much, the diffraction grating will cause the image of the field diaphragm aperture to appear with a halo around it. This aperture halo will be superimposed on the specimen. Too much light, on the other hand, may cause glare and impair the resolution of the viewed image.

Step 10. Remove the eyepiece and examine the back lens of the objective. Adjust the microscope substage iris diaphragm so that the transmitted light almost fills the objective lens.

Step 11. If the light passing through the microscope and reaching the eye is too intense for comfortable viewing, decrease the intensity by placing a neutral density filter in front of the illuminator. Do not use the substage iris to cut down the light.

Replacing one phase objective with another will probably not radically change the performance of the phase-contrast microscope. When changing slides, however, it will generally be necessary to adjust the focus of the substage condenser.

Phase-Contrast Microscopy of Opaque Objects Using Incident or Reflected Light

Phase-contrast microscopy can also be used for opaque objects by illuminating the specimens with incident light through the use of a vertical illuminator (see page 277).When the specimen is viewed with incident light, it must be exactly perpendicular to the optical axis of the microscope, since the annular aperture diaphragm is also centered on this optical axis.

If the light striking the opaque specimen is too diffused, no image of the aperture of the annular diaphragm is formed and consequently no phase contrast of the specimen is achieved. The amount of phase contrast possible with opaque objects using reflected light depends largely on the irregularities of the specimen. Bright or dark contrast causes the depressions of the specimen to appear brighter or darker, respectively, against the surrounding area.

General Rules for Obtaining Good Phase Contrast

Several factors that will aid the microscopist in achieving good phase contrast in specimens are discussed in the following paragraphs. Living specimens such as living tissues or live unicellular organisms are imbedded in water, nutrient liquid, or isotonic media. Whatever the medium, it must be nontoxic and must not adversely affect or alter the specimen. Ideally, it should be practically colorless and it should neither evaporate too rapidly nor deteriorate with time. It is also important that the chosen medium have a different index of refraction than the specimen. For nonliving specimens, various media are tried until one is found that gives the desired results. Changing the temperature of the medium by using a warm stage will sometimes change its index of refraction and thus give different views of the specimen. If the specimen shows too rapid movement, it should be mounted in a more viscous medium to slow or stop motion.

Microscope glass slides about 1.2 mm thick are generally used for the standard phase-contrast microscope. Cover glasses about 0.18 mm thick are used with these slides, since most phase objectives are corrected for this thickness. The cover glass is placed over the specimen and sealed with petrolatum to prevent evaporation. Hanging drop mounts and cell mounts are not suitable for phase-contrast microscopy, since the drop or cell mount acts as an auxiliary lens, causing some deflection of the light and preventing the exact focusing of the aperture of the annular diaphragm within the conjugate area of the diffraction grating.

Any specimens chosen for examination should be thin and uniform so that they are transparent and do not hinder the transmission of light.

The correct combination of annular aperture diaphragm and phase objective should be used, so that the image of the annulus will coincide with the image of the conjugate area. Köhler illumination gives best results.

Lightly stained specimens may show more details, enhance visibility, and give better contrast with the phase-contrast microscope. It may prove helpful to view the same specimen under bright and dark contrasts, especially if the specimen is a two-phased emulsion.

When viewing colored specimens, color filters may increase visibility. Normally, a complementary colored filter is placed between the microscope illuminator and the microscope.   It will usually increase contrast.   At times, a filter of a color similar to that of the specimen may reveal additional details; thus, it is advisable to try both types of filters.

Anoptral System of Phase-Contrast Microscopy

Alvar Wilska found that another type of objective useful in phase-contrast microscopy can be prepared by applying a coating of soot from a candle flame.   The soot is deposited on the upper objective lens in such a way as to form an absorbent soot coating, recessed in the form of a ring with a clear annular opening or aperture. A matching annular diaphragm placed beneath the condenser produces phase-contrast images that are practically free from bright halos and that give a slight illusion of depth while forming a golden-brown tinted image.   This phenomenon results from the light being absorbed by the soot layer instead of being reflected, as is the case with ordinary phase-contrast objectives.   This system of phase-contrast microscopy, using a heavily sooted absorbing annulus, is called *anoptral contrast*. Commercially, soot has been replaced with other materials, and this type of phase objective is sold under the name of *anoptral contrast objective*.

CHAPTER 9

The Interference Microscope

THE TWO-BEAM INTERFERENCE MICROSCOPE (see Fig. 121), which is some-

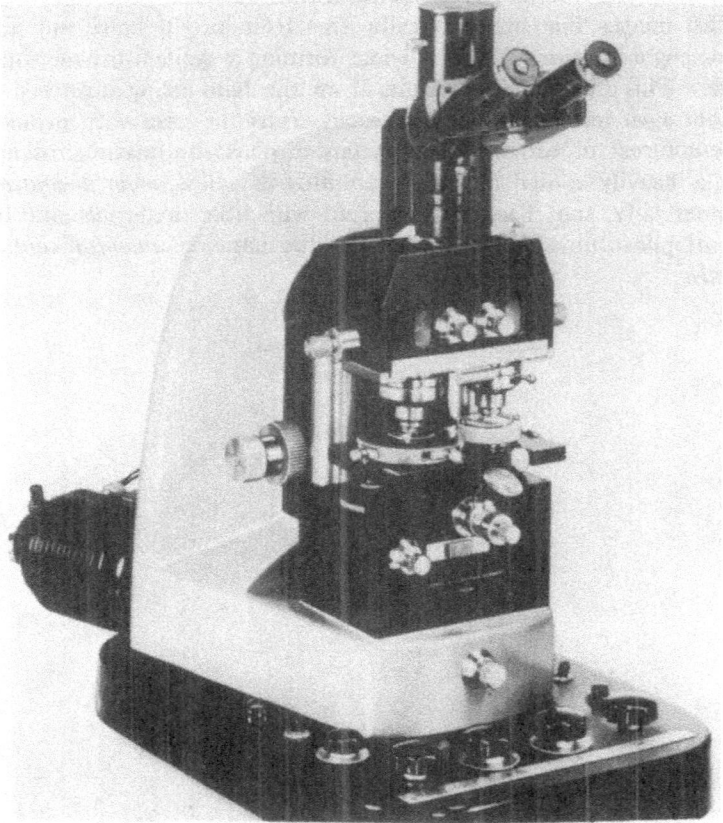

Fig. 121. Two-beam interference microscope.

234

what similar to the phase-contrast microscope, depends on the concept of separating or splitting a single beam of transmitted light into two beams. One beam, called the *object beam,* traverses the specimen and is thereby deviated or retarded relative to the second beam, called the *reference beam,* which proceeds along one side of the specimen, without contacting it. The two beams are then recombined by means of prisms; at this point, interference of the light waves occurs, with the result that the contrast of the specimen is increased. If white light is used, details within the specimen are seen in variable colors; if monochromatic light is used, the intensity of contrast is increased.

The image produced by a two-beam interference microscope is substantially free of halos. Moreover, the microscope reveals certain interference patterns that can be measured, making it possible to determine the exact optical path difference. Since the optical path of a specimen is the product of its thickness and its index of refraction, any one of these three variables can be easily determined with the interference microscope if the other two are known. Thus, the interference microscope is primarily a measuring instrument.

GENERAL PRINCIPLES

Essentially, the interference microscope is two instruments—an interferometer and a microscope. The basic concept of the interferometer is illustrated in Fig. 122.

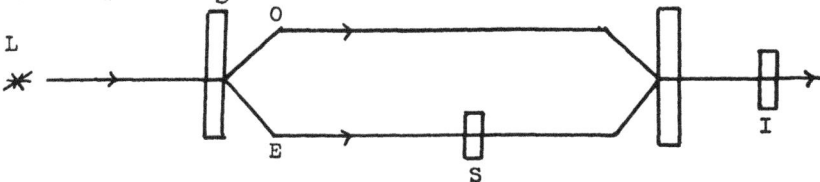

Fig. 122. Diagrammatic representation of the concept of the interferometer (discussed in the text).

Light from a source, L, strikes a birefringent crystal, P. The light is split into two beams, one of which, O, travels without hindrance. The second beam, E, passes through the dense specimen and is retarded relative to O, which is passing through the rarer medium. At P', another birefringent crystal similar to P, the two beams, O and E, are recombined to form the image at I. Since O and E are out of phase, interference of the light occurs and an increase in contrast is produced.

There are many methods of obtaining microimages by the split-beam interference procedure. Most modern interference microscopes use polarizers and refracting plates to accomplish this.

METHODS OF MEASURING PHASE DIFFERENCE

If a blank slide is placed on the stage of the two-beam interference microscope and viewed with white illumination, light and dark bands or fringes are seen. If monochromatic light, say green, is used, a pattern of parallel dark bands appears against the green background. The substage condenser can be adjusted so that the width of one dark fringe is increased to cover the field of view with uniform intensity. At this point, there is practically no interference between the two beams. If a specimen is now placed on the slide in the path of the object beam, this beam will be deviated as it passes through the specimen, and interference will result when the object and reference beams recombine. The effect of this phenomenon as one looks through the microscope, is to brighten the field of view. This increase in brightness is produced by the phase difference of the retarded object beam coming through the specimen and the virtually unobstructed reference beam coming through the surrounding medium. The change in intensity, or constructive phase difference, of the light can be measured with an interference microscope such as the Baker type, described below.

Another method of measuring phase difference is to use white light. If an empty field is viewed with a two-beam interference microscope, a number of parallel, evenly spaced light- and dark-colored bands are seen. If a phase-retarding plate or wedge compensator is placed in the path of the reference beam and a specimen in the path of the object beam, the light- and dark-colored bands are displaced from their original positions and are clearly visible (see Fig. 123). The degree of displacement or the numerical magnitude of the displaced fringe, which depends on the thickness and refractive index of the specimen,

Fig. 123. Diagrammatic representation of the displacement of interference bands (discussed in the text).

can be measured. This measurement is made with another similar, movable wedge compensator attached to a graduated drum, on the scale of which the amount of displacement can be read. The Leitz two-beam interference microscope utilizes this method of measuring phase difference.

Knowledge of the polarizing microscope, which is discussed in Chapter 7, is helpful in understanding the interference microscope.

Various interference microscopes are sold, each designed somewhat differently to achieve interferometric effects. We will describe these designs in general, and the Baker type in some detail. Immediately following, however, is a description of the multiple-beam method of obtaining interference effects. This method can be easily applied to the standard microscope for examining transparent and opaque specimens and for measuring optical path differences, although the split-beam interference microscope is more effective for measuring transparent biological specimens.

Multiple-Beam Method for Transparent Specimens

The transparent specimen is mounted on a glass slide that has been partially coated with a highly reflective metallized material and covered with a cover glass coated with the same type of material. Monochromatic light, generally green, is passed through a 0.5-mm pinhole, which is positioned below the condenser at its focal point, to collimate the light. The light waves strike each separate portion of the specimen and are repeatedly reflected because of the metallized surfaces of the slide and cover glass. As one looks through the microscope, fringes produced by the interference of this light are visible.

If the phase difference of the transmitted light waves is one wavelength, constructive interference occurs and a bright image is visible. The distance between every two bright fringes, therefore, is one wavelength. Each single fringe has the appearance of a contour line on a map, delineating the light waves with equal paths. Any deformation of one or more of these fringes caused by the interference or phase shift of the light passing through the specimen can be used to determine the optical path difference. The linear distance by which the fringes are deformed or displaced is a measure of the interference introduced by the parts of the transparent specimen. The method works rather well if the difference between the refractive index of the specimen and that of the surrounding medium is small; otherwise, unwanted reflections may occur.

TWO-BEAM INTERFERENCE MICROSCOPE

Leitz Interference Microscope (Mach-Zehnder Type)

With this instrument, the light beam is split into two parts—the object beam and the reference beam. Complete separation of the object beam, which passes through the specimen, and the reference beam, which does not, is accomplished by passing the two beams through separate condensers and objectives. Actually, the system incorporates two microscopes, positioned 62 mm apart.

The specimen slide is placed on one stage, a comparison slide on the second stage. Two identical microscopic objective are used, one to receive light transmitted through the comparison slide, the other to receive light transmitted through the specimen.

The Leitz interference microscope, which is shown in cross section in Fig. 124A, consists essentially of a lower prism housing, an upper prism housing, and the optical system. The lower prism housing, T_1, which is positioned below the two objectives, O, contains the beam-splitting prisms, which transmit the object beam and the reference beam through separate compensator plates. The two beams are roughly centered by means of a condenser centering mechanism, K. Final adjustment is made with the aid of an additional rotating plate compensator, P_1. After adjustment, the interference bands seen through the microscope will appear in intense contrast against the surrounding background.

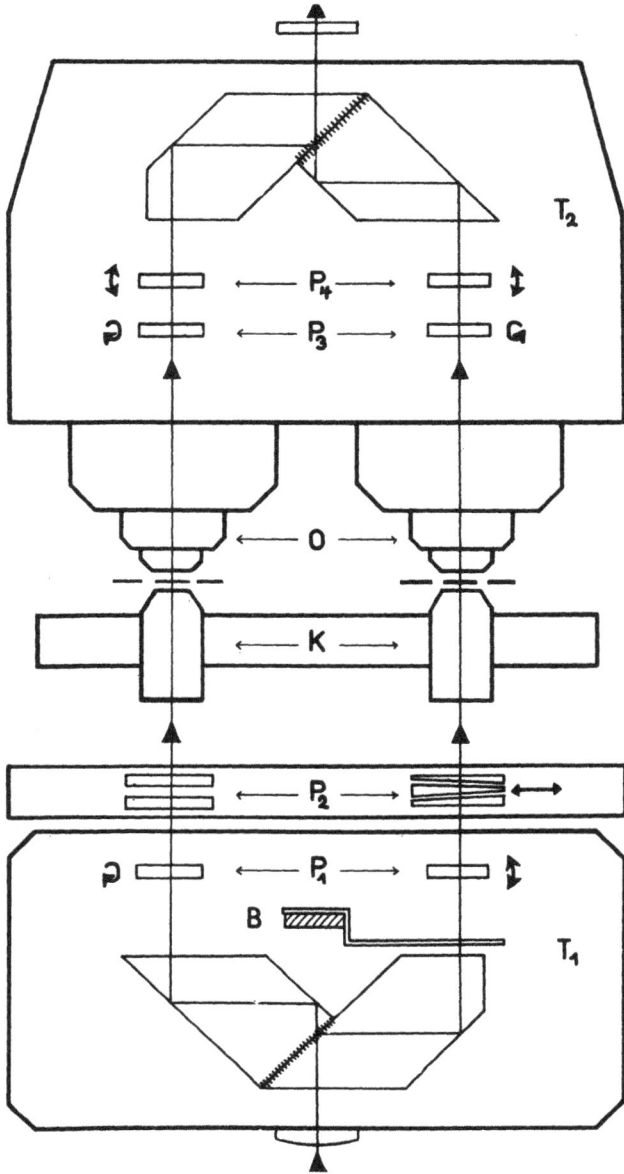

'g. 124A.   Cross section of the Leitz interference microscope (described in the text).

The lower housing also contains a wedge compensator P_2, posi-- tioned below the reference slide, and used to determine the optical path or phase difference, which is indicated on an attached graduated measuring drum (1 division=4.0° rotation at 546 nanometers). A more detailed view of this device, also in cross section, is provided by Fig. 124B.

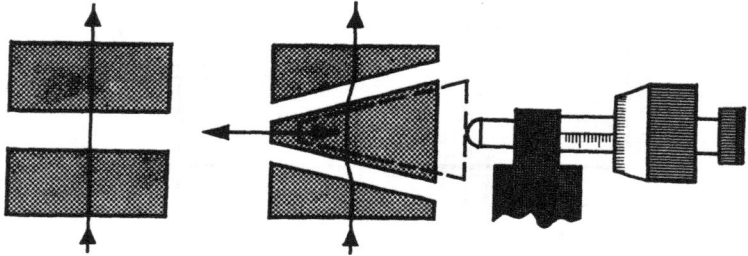

Fig. 124B.   Cross section of the graduated wedge compensator.

Returning to Fig. 124A, the upper prism housing, T_2, positioned above the objectives, contains two pairs of plate compensators, P_3, P_4, which can also be tilted to orient and adjust the interference fringes of the light. The two beams are reunited before entering the body tube of the microscope by an upper prism positioned above these com- pensators.

The advantage of this system is that it prevents the reference beam from passing through the specimen or any part of the specimen slide, since the path of the object and reference beams are so far apart. Under these circumstances, there is little limit as to the size of the specimen that can be handled properly.

Baker Interference Microscope (American Optical)

The *Baker interference microscope,* made by American Optical, is basically a polarizing microscope that uses a birefringent calcite plate positioned above the condenser to split the plane-polarized light into two beams—the ordinary and the extraordinary beam (see page 188).

The instrument is available with two interference systems, called *shearing* and *double focus*, the only difference being in the orientation of the optic axes of the birefringent calcite plates. In the shearing system, the optic axes are inclined at an angle of 45° to the faces of the plates; in the double-focus system, the optic axes are parallel to the faces (see Figure 124C).

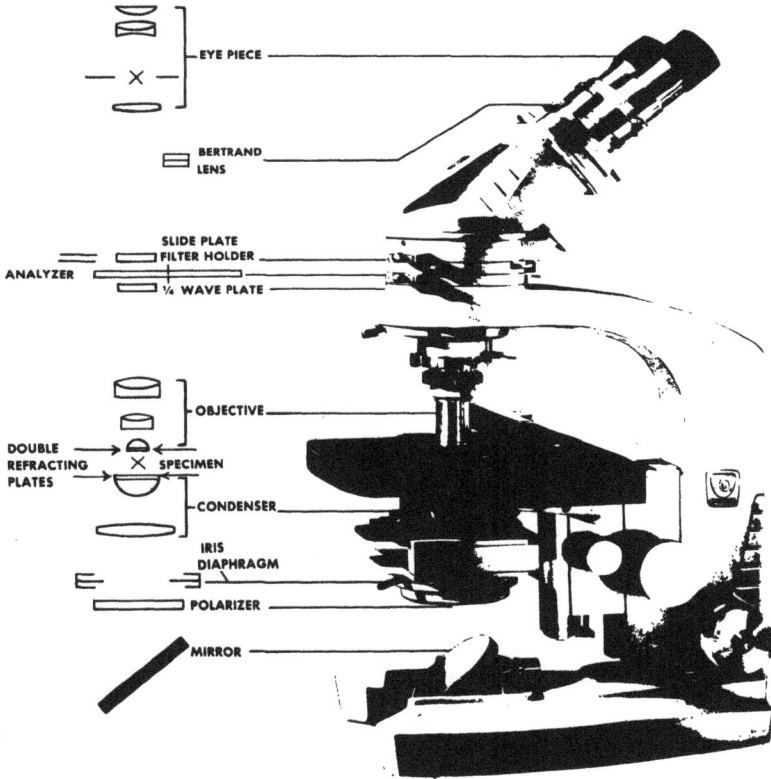

Fig. 124C. Interference microscope.

SHEARING SYSTEM

When the shearing system is used, the path of the beam of light passing through the Baker interference microscope is as follows: The light strikes the mirror and is reflected upward through the polarizer, where it is plane-polarized at 45° to the optic axes of the two double-refracting calcite plates.   The polarized light passes through the substage iris diaphragm and then through the condenser, where it encounters the first birefringent calcite plate, which is positioned above the condenser.   Since the optic axis of the calcite plate is inclined at an angle of 45° to its face, the extraordinary, or object, ray is refracted, or "sheared" to the left and passes through the specimen positioned on the glass slide; the ordinary, or reference, ray passes to one side of the specimen and continues unhindered through the blank portion of the slide (see Fig. 125).

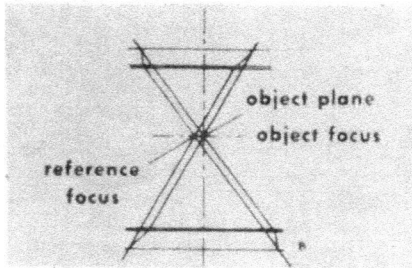

Fig. 125. Diagrammatic representation of the path of light through
the shearing system (described in
the text).

The two separated rays, which are traveling along a common path, encounter another half-wave plate.   As the light passes through this plate, the plane of polarization is rotated 90°, which means that the extraordinary ray becomes the ordinary and vice versa. The rays are then reunited by the second birefringent calcite plate, which is of the same thickness and orientation as the calcite plate that originally separated the light.   The two beams are now traveling together through the objective along a common ray path.

Normally, there is a phase difference between the reference beam and the object beam that causes constructive interference, so that more details of the specimen become visible. This phase difference represents the optical path of the object beam passing through the specimen minus the optical path of the reference beam passing to one side of the specimen.

To detect and measure this phase difference exactly, a quarter-wave length retardation plate is placed above the objective and positioned so that its optic axis is at a 45° angle to the plane of vibration of the object beam. The reference and object beams, traveling along common paths, pass through the quarter-wavelength plate and are converted into circular polarized light, one beam rotating clockwise, the other counterclockwise. The result of these two opposite circularly polarized beams is merely plane-polarized light, the direction of which depends on the optical path difference between the two beams. Consequently, it is possible, by means of an analyzer placed above the quarter-wavelength plate, to determine the optical path difference by turning the analyzer until the *surrounding area* close to the specimen appears darkest. This position of the analyzer is noted in degrees on the analyzer scale. The analyzer is again rotated until the *specimen proper* appears darkest, and this position is noted on the analyzer. The difference between these two settings is half the phase difference introduced by the specimen; the actual path difference, in degrees, is therefore twice this measured angle.

DOUBLE-FOCUS SYSTEM

As noted above, the only difference between the shearing and the double-focus systems is the orientation of the two birefringent calcite plates. In the latter system, plates are positioned so that their optic axes are parallel to their faces. The object beam focuses at the specimen, while the reference beam spreads around the specimen and is brought to a focus directly above it. The separated object and reference beams are then recombined in a manner similar to that of the shearing system (see Fig. 126).

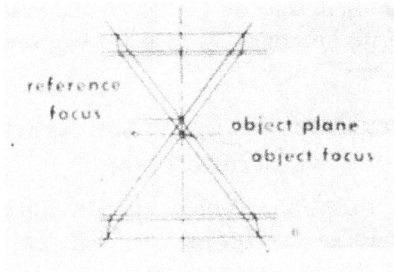

Fig. 126. *Diagrammatic representation of the path of light through the double-focus system (described in the text).*

ADVANTAGES AND DISADVANTAGES OF THE TWO SYSTEMS

The double-focus system is well adapted to the qualitative examination of specimens where exact measurements are not required. It gives better visual contrast, since the reference beam is less affected by the surrounding medium. A notable disadvantage, however, is that very small specimens must be used to avoid overlapping between the object beam, focusing at the specimen, and the reference beam. The size of the specimen should not exceed 95 micrometers for the 10X and 40X double-focus objectives or 25 micrometers for the oil-immersion double-focus objective.

The shearing system is especially useful for measuring the features of a specimen, since there is complete lateral separation of the object and reference beams. There is also a limit to the maximum size of the specimen in this system, although it is much greater than in the double-focus system and there is no overlap between the object and reference areas. The size should not exceed 330 micrometers in diameter for the 10X and 40X shearing objectives or 160 micrometers for the oil-immersion shearing objective. If overlapping does occur, the specimen must be decreased in size so that the reference beam is not obstructed as it passes through the microscope.

Overlapping of the specimen can be determined by looking through the microscope as the specimen is brought to a focus. On the left side of a proper-sized specimen, a blurred image of this specimen will appear. The extent of this area of the focused specimen as compared to the blurred image actually determines the maximum size of the specimen that can be observed with the shearing system; therefore an equal area on the right side of the specimen must be clear and free from any part of the specimen. If it is too big, the specimen must be brought to this size.

General Procedures for Using the Baker Interference Microscope

The instructions that follow, which explain specifically how to use the Baker interference microscope, are also generally applicable to other types of interference microscopes.

ILLUMINATION

For general observation, a tungsten microscope lamp equipped with a condenser and iris diaphragm is adequate. White light is useful in bringing out color contrasts within the specimen. For quantitative measurements such as optical path differences, however, coherent illumination is essential.

COHERENT LIGHT

There are two elements necessary for coherent light—*temporal coherence* and *spatial coherence*. Temporal coherence is achieved by using monochromatic light having a very narrow spectral range. Temporal coherent light is essential when the interference microscope is used as a measuring instrument, since light waves coming from this source of illumination are of a single frequency and produce a single pure color. Spatially coherent light waves, which are derived from a single point source, are in phase and can therefore be imaged as a small spot or point. Such beams show minimum divergence in their passage and are highly directional.

A mercury arc lamp used as a source of illumination meets the requirement for temporal coherence. Monochromatic light is obtained by passing the light from the mercury arc lamp through a narrow-band filter, a procedure that eliminates all spectral components except those in a very narrow spectral range.

Spatial coherence is partially obtained in the interference microscope by passing monochromatic light through a polarizer and then through a small aperture of the iris diaphragm. The light is then collimated and split by means of birefringent plates.

Green monochromatic radiation with a wavelength of 546 nanometers is generally chosen because the birefringent calcite plate and the quarter-wavelength retardation plate are standardized for this radiation, because the eye is most sensitive to this wavelength, and, finally, because the actual wavelength of the monochromatic light must be known for any measurment of the optical path. Mercury arc lamps emit green radiation of this wavelength. To further ensure that the green light transmitted through the interference microscope is isolated and almost entirely of 546-nanometer wavelength, special filters are used with the lamp to exclude most other such minor emissions as yellow and blue. Note that filters diminish the intensity of the transmitted light somewhat. Table 5 lists a few of the available filters, together with their characteristics, that are suitable for use with the mercury arc lamp.

TABLE 5. FILTERS FOR USE WITH THE MERCURY ARC LAMP

Filter Number	Principal Lines Transmitted (nanometers)
Wratten 22	577, 578 (yellow), 71%
50	436 (blue), 6.5%; some 408, 405, 398
74	546 (green), 10%; 577, 0.2%
77	546 (green), 74%, 577, 0.5%
77A	546 (green), 68%; 577, negligible
18A	310–390, isolation of 365 (ultraviolet), 60%
58	546 (green), 42%; 577, 11%
77+58	546 (green), 31%, used to eliminate most red lines
77A+58	546 (green), 29%; used to eliminate all red lines
Corning CS4-120	546 (green), 44%; no blue or yellow lines
CS-584	436 (blue), 22%; no green or yellow lines

Köhler illumination is generally used with the interference microscope for visual study and photomicrography. The filaments of the tungsten or mercury arc lamp are exactly centered in the microscope and any filters or diffusing glass plates are temporarily removed. Set the lamp from 5 to 7 in. away from the microscope, or in a position where the image of the lamp or the outline of the arc fills the opening of the condenser sufficiently. Continue with the procedure for Köhler illumination described on page 42.

PREPARING THE SPECIMEN

The preparation of a specimen for study under the two-beam interference microscope is critical, especially when quantitative measurements are to be made. The slides and cover glasses chosen must be thoroughly clean and especially free of scratches, waviness, or other defects. The thickness of the cover glass should be about 0.18 mm (No. 1½), since most interference objectives are adjusted for this thickness. The liquid mounting medium must be filtered or centrifuged so that it is homogeneous and pure. Any adulterants in the liquid medium or defects on the slide and cover glass will lead to errors in computing phase differences.

ADJUSTING THE MICROSCOPE

Each objective for the interference microscope is supplied with a matched condenser. It is therefore essential that the objective and its corresponding condenser be properly and securely mounted in the microscope.

PROCEDURE

Step 1.   Place the specimen on the stage.   Illuminate it and bring it into focus.

Step 2.   Rotate the control lever of the polarizer located above the mirror to the position marked INT. and insert the quarter-wavelength retardation plate.

Step 3.   For the shearing system, rotate the analyzer to the setting indicated on the manufacturer's reference card that comes with the shearing objective and condenser. For the double-focus system, rotate the analyzer to about 100°, which positions the analyzer in the light path.

Step 4.   Remove the eyepiece and replace it with a telescope provided with a Bertrand lens.   (This device is similar to the one used in the polarizing microscope for focusing interference figures.)

Step 5.   Replace the ground glass in the microscope lamp to diffuse the light, then open the substage iris diaphragm so that the condenser is adequately filled with light.   Looking through the telescope, move the slide so that a clear area of the slide, not the specimen, is in the field of view. A pattern of alternating uniform bright and dark bands or fringes of light will then be seen across the back opening of the objective.   For optimum results, these bands must be adjusted. Select one of the dark bands near the central portion of the field of view and spread it uniformly across almost the entire field by turning the tilting screws located on the condenser mount (see Fig. 127).   If

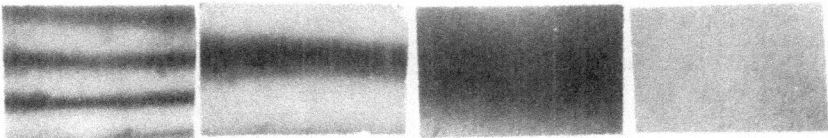

Fig. 127.   Photomicrographs showing the procedure for centering and spreading a dark band (discussed in the text).

this dark band does not spread uniformly, try another near the center portion of the field.

If shearing objectives and condensers are used, close down the substage iris until two thirds of the field is uniformly filled with the bandspread. If double-focus objectives and condensers are used, do not close down the iris; the condenser must be fully open to allow passage of the reference beam, since otherwise the beam might be cut off.

Step 6. Remove the telescope and return the eyepiece. Move the specimen into the optical path and bring it into sharp focus. If white light is now used, the specimen is seen in color that can be made to change by rotating the analyzer.

When using the oil immersion objective and its companion condenser, *do not* spread the immersion oil on the condenser in such a way as to form an oil film between it and the slide.

Making Quantitative Measurements with the Baker Interference Microscope

DETERMINING THE OPTICAL PATH DIFFERENCE

The shearing system is preferred for determining optical path differences of details within a specimen or for other quantitative measurements. The optical path difference between the specimen and its surrounding medium can be expressed mathematically by the equation:

$$\text{OPD} = \frac{2(\theta_1 - \theta_2)}{360} \lambda \tag{1}$$

in which OPD is the optical path difference; θ_1, θ_2 are the readings of angles for the background and the specimen, respectively, when the analyzer is rotated to achieve minimum brightness of each; and λ is the wavelength of the monochromatic light used (usually green light with a wavelength of 546 nanometers).

Monochromatic light of definite, known wavelength must be used to determine the optical path difference. A mercury arc illuminator is used, and the green monochromatic light transmitted from it (546 nanometers) is isolated by the appropriate filter (see Table 5, page 247).

One full rotation of the analyzer (360°) corresponds to one wavelength of phase difference. The Baker interference microscope can measure the optical path difference with great accuracy only when this difference is less than one wavelength. When it is greater, unwanted interferences are encountered. Consequently, a mounting medium with a refractive index close to that of the specimen must be used.

CORRECT ROTATION OF THE ANALYZER

In determining the extinction position of the specimen and background, the analyzer must be rotated in the correct direction; if it is not, an error of $\pm 180°$ may occur. To determine whether the analyzer is to be turned clockwise or counterclockwise, one must know whether the refractive index of the specimen is greater or less than that of the surrounding medium.

The Becke line can be used to make this determination for the detail of a specimen. Turn the polarizer to the OFF position, remove the quarter-wavelength plate, and determine the Becke line as described on page 197.

The direction (as one looks down through the eye tube) in which the analyzer should be rotated to go from the extinction position of the surrounding medium to that of the specimen, using either a shearing or a double-focus objective, is determined by the following rules:

For *10X and 40X objectives of both types* and the *100X shearing objective:* If the refractive index of the specimen is *less* than that of the medium, all turns are *clockwise;* if the refractive index of the specimen is *greater* than that of the medium, all turns are *counterclockwise.*

For the *100X double-focus objective:*   These directions are *reversed.*

EXTINCTION TRANSFER METHOD

Illuminate the specimen with green monochromatic light. Rotate the analyzer until a clear region, as close as possible to the detail of the specimen, appears darkest. Read the position of the analyzer scale in degrees. Rotate the analyzer again in the correct direction until the detail is made as dark as possible. Read this position of the analyzer scale in degrees. The difference in degrees between these two settings is half the optical path difference introduced by the specimen.

CALCULATING THE OPTICAL PATH DIFFERENCE

For example, assume that the optical path difference of a cell nucleus mounted in a water medium is to be determined. A 40X shearing objective and its companion condenser are mounted in the microscope. The refractive index of the cell nucleus will normally be slightly greater than that of the surrounding water medium. The extinction position of the water medium read on the analyzer scale is found to be 162.4°. Since the refractive index of the nucleus is greater than that of the surrounding medium, the analyzer is rotated counter-clockwise, following the rules stated above, and the extinction position of the nucleus as read on the analyzer scale is found to be 213.4°.

Using green monochromatic light (wavelength 546 nanometers) and substituting its wavelength and the values for the two extinction positions in Eq. (1), we find the optical path difference to be :

$$OPD = \frac{2\,(213.4 - 162.4)}{360} \times 546 = 154.7 \text{ nanometers}$$

MATCHING METHOD

The matching method can be used when it is known that the greatest optical path difference between any two parts or details of the specimen is not more than half a wavelength of the monochromatic light used.

Rotate the analyzer until the surrounding medium close to the specimen is at minimum brightness. Record this reading in degrees from the analyzer scale. Rotate the analyzer in the correct direction until the outer area of the specimen and the background surround appear equally bright or dark. The purpose is to match the lighting intensities of the specimen area and the background medium area. Record this analyzer scale reading. The difference in these analyzer readings, multiplied by four, is the optical path difference, in degrees, of the specimen portion being measured.

The matching procedure can be continued by matching internal details of the specimen, such as the nucleus, with reference to the outer measured portion of the specimen as explained above.

The sum of the optical path difference of an internal portion of the specimen, such as the nucleus, and the optical difference of the marginal portion of the specimen, the cytoplasm, will give the total optical path difference of the nucleus alone when referred to the surrounding medium of the specimen. Care must be taken that the reference areas are clear, clean, homogeneous, and free from variations.

Results obtained with the matching method and the extinction transfer method are normally in very close agreement.

HALF-SHADE EYEPIECE

The *half-shade eyepiece* (see Fig. 128), although it is a rather expensive accessory, permits greater precision when matching the brightness of the light in two adjacent areas. The eye is more sensitive when matching the luminance in adjacent areas than when determining the difference in brightness of large isolated areas. The half-shade eyepiece, which is designed for use with the monocular interference microscope, exploits this sensitivity.

The eyepiece, which has its own analyzer with a scale graduated in tenths of a degree, outlines a field that is exactly divided into two. To determine optical path differences with the half-shade eyepiece, turn the microscope analyzer to the OUT position and insert the quarter-wavelength plate in the optical path of the microscope. Center the specimen on the stage so its image is directly beneath the dividing line of the half-shade eyepiece and rotate the eyepiece analyzer until there is equal darkness on both sides of the dividing line in the eyepiece. Record the reading on the half-shade analyzer scale. Move the specimen until the background area adjacent to the specimen is centered beneath the dividing line. Rotate the eyepiece analyzer until there is again equal darkness on both sides of the dividing line. Record the second reading on the analyzer scale. The optical path difference is computed in the same way as in the extinction-transfer method.

QUARTZ-WEDGE EYEPIECE

The Baker interference microscope can be used only when the optical path difference between the specimen and its surrounding medium is less than one wavelength of the monochromatic light used. If this difference is greater than one wavelength, the fractional part of the wavelength is first measured with the interference microscope. The remaining number of full wavelengths of the path difference can be determined with the *quartz-wedge eyepiece* (see Fig. 129). The quartz compensator, which is part of the eyepiece, helps in determining the

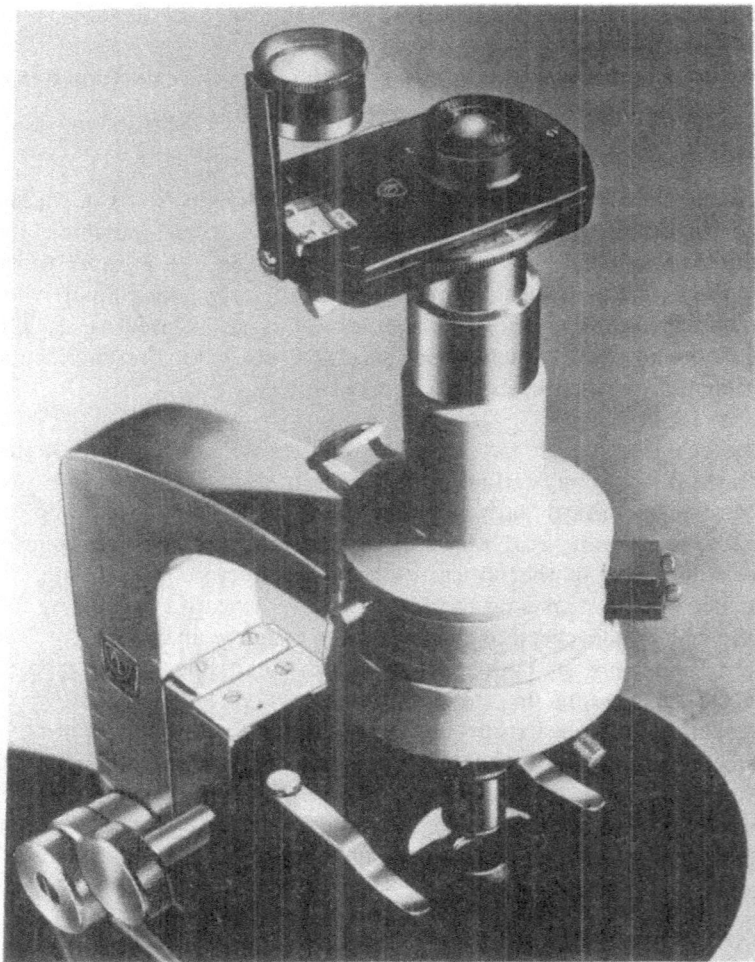

correct order number of the interference colors observed in the specimen. Thus, to determine the number of full wavelengths or, technically speaking, the order of interference colors, this eyepiece is used in place of the regular one.

Fig. 129.   Quartz-wedge eyepiece.

PROCEDURE

Turn the analyzer and the quarter-wavelength plate to the OUT position. White light is used to illuminate the specimen instead of monochromatic light. The slow direction of the quartz, which is engraved on the wedge, is aligned either parallel to or at right angles to the microscope arm, which is in the plane of the optical axis of the microscope. For specimens with a greater refractive index than their surround (the normal case), the slow direction of the wedge is aligned parallel to the support arm.

As the wedge is moved across the field, colored fringes are visible, one of them black. Move the specimen slide until the black fringe is positioned in the surround near the specimen; then move the wedge until the black fringe is opposite or in line with the specimen. Count the number of colored fringes that pass through the specimen during the movement of the black fringe from the surround until it is in line with the specimen. This number is equal to the number of full wavelengths of the optical path difference of the specimen relative to its surround.

CALCULATING THE THICKNESS AND REFRACTIVE INDEX FROM THE OPD

There is a relationship among the optical path difference (OPD) of the specimen, its refractive index, and the refractive index of its surrounding medium that can be expressed mathematically by the equation:

$$\phi = t\,(n_2 - n_1) \tag{2}$$

in which ϕ is the optical path difference; t is the thickness of the specimen; n_2 is the refractive index of the specimen; and n_1 is the refractive index of the medium through which the reference beam passes, generally the mounting medium.

For example, say that the thickness of the specimen, in a water mount, is found to be 1.5 micrometers. The optical path difference of the specimen, using green monochromatic light, is found to be 0.1 micrometer. The refractive index of the specimen, or of the mounting medium, can be easily ascertained from the Eq. (2). If the refractive index of the mounting medium is known (in this case, water, with a refractive index of 1.33), the refractive index of the specimen can be calculated:

$$0.1 = 1.5\,(n_2 - 1.33) \tag{2}$$
$$n_2 = \frac{1.995 + 0.1}{1.5}$$
$$n_2 = 1.3967$$

If the specimen is immersed successively in two different media of known refractive index and the optical path difference of the specimen is determined in each medium, both the refractive index and the thickness of the specimen can then be calculated at the same time. The relationships for each medium can be expressed by Eq. (2) and a variant in which ϕ' and n'_1 represent ϕ and n_1 for the second medium:

In one medium: $\phi = t\,(n_2 - n_1)$ $\tag{2}$

In second medium: $\phi' = t\,(n_2 - n'_1)$ $\tag{3}$

Subtracting Eq. (3) from Eq. (2) gives Eq. (4) below, which can be used to determine the thickness of the specimen; dividing Eq. (2) by Eq. (3) gives Eq. (5) below, which can be used to determine the refractive index of the specimen:

$$t = \frac{\phi - \phi'}{n'_1 - n_1} \tag{4}$$

$$n_2 = \frac{\phi n'_1 - \phi' n_1}{\phi - \phi'} \tag{5}$$

A highly sensitive method of determining the refractive index of the specimen is to change its surrounding medium until the known refractive index of the medium is equal to that of the specimen. Under these circumstances, there will be no interference of the light, and the optical path difference will be zero.

CALCULATING THE MASS OF AN ORGANIC SPECIMEN

The interference microscope can be used advantageously to determine the dry mass of single cells or cell components found in dilute aqueous solutions. The refractive index of a dilute water solution of a starch, fat, protein, or other organic material increases directly with the concentration of the substance in the solution. When concentration is plotted against refractive index on a graph, a straight-line graph results and the specific refractive increment is found to be uniform for each material used. This increment can be expressed mathematically by the equation:

$$\alpha = \frac{n_s - n_w}{C} \tag{6}$$

in which α is the specific refractive increment which varies little in many biological substances; n_s is the refractive index of the solution; n_w is the refractive index of water (1.33); and C is the concentration of the dry material in g/100 ml of solution.

For proteins (protoplasm), the average value of $\alpha = 0.0018$; for fats, $\alpha = 0.0014$; for starch, $\alpha = 0.0013$; and for nucleic acids, $\alpha = 0.0016$.

If a specimen cell is mounted in water, the optical path difference can be similarly expressed as:

$$\phi = t(n_s - n_w) \tag{7}$$

or

$$\frac{\phi}{t} = (n_s - n_w) \tag{8}$$

Substituting $\dfrac{\phi}{t}$ for $n_s - n_w$ in Eq. (6), we have:

$$\frac{\phi}{\alpha} = Ct \tag{9}$$

The quantity Ct is considered the dry mass per unit area of the specimen. It therefore follows from Eq. (9) that:

$$M \text{ (dry mass)} = Ct = \frac{\phi}{\alpha}. \tag{10},(11)$$

To determine the dry mass of an entire specimen, Eqs. (10) and (11) become:

$$M = \frac{ACt}{100} = \frac{A\phi}{100\alpha} \tag{12},(13)$$

in which A is the area of the specimen in sq. cm.

To determine the quantity t for use in Eqs. (12) and (13), use Eq. (2). When t is known, C is easily determined from Eq. (9).

If the refractive index of the specimen is determined by Eq. (5), this value can then be used to determine C in Eq. (6).

For an example that shows the determination of the mass of a microscopic specimen by using Eq. (13), assume that the nucleus of a protozoan, which is circular, has an average diameter of 22.4 micrometers. The area of the nucleus in square centimeters will be πr^2 $= 3.142 \ (11.2 \times 10^{-4})(11.2 \times 10^{-4})$, or $A = 394 \times 10^{-8}$ sq. cm.

The determined path difference of the nucleus is 0.16 micrometer (16×10^{-6} cm). The value of α for protein or protoplasm is 0.0018. The mass of the nucleus, calculated by substituting these values in Eq. (13), would be:

$$M = \frac{(394 \times 10^{-8})(16 \times 10^{-6})}{100 \times 0.0018} = 349 \times 10^{-12} \text{ g.}$$

Precautions in the Use of the Interference Microscope

The following factors should be considered when using the interference microscope :

1. Since the optical paths of the object and reference beams are determined by the irregularities in their paths, these two beams will cancel each other out if the irregularities are the same in both paths. To make sure that the optical path readings are due only to the specimen and the surround, it is essential that the following precautions be adhered to: The slides, cover glasses, mirror, and optical parts must be thoroughly clean, more so than in ordinary use; (b) the glass slide, the cover glass, and the mounting medium must not be tilted but must be kept exactly parallel to the stage, so that no deviation in the optical paths of the beams is caused by any wedge effects in the specimen slide (to ensure that this non-parallelism does not occur, rotate the stage and observe whether there is any change in the brightness of the object or the background); (c) the mounting medium must be transparent, homogeneous, and free of any particles.

2. The observed specimen should be thin and of uniform thickness. Thick or curved specimens will cause variations in the optical paths within the specimen and consequent errors in measurements.

3. The temperature of the specimen and the mounting materials should be kept as close as possible to room temperature. The refractive index of liquids varies with changes in temperature, causing errors in the measurement of optical path differences.

4. The optics used in the interference microscope must be strain-free.

5. If precise measurements are to be carried out, the correct selective filter must be used to give the exact wavelength of monochromatic light required.

6. The light from the illuminator must be carefully centered in the mirror so that the light travels through the microscope directly, not obliquely. To minimize this type of error, close the condenser

down to a numerical aperture of about 0.6 or slightly less (closing down the condenser much further would result in a loss in resolution).

7. Do not move the centering screws after adjusting and centering the objectives.

8. When birefringent specimens are examined, the optic axis should be aligned parallel to the plane of the polarized light; otherwise, errors in measurements might occur. Organic specimens are only slightly birefringent and therefore do not cause any appreciable errors.

9. When setting the analyzer for the position of maximum darkness for the specimen and its surround, it is essential that it be set at the extinction position and not at the maximum contrast of the specimen against its background. Avoid glare and too intense illumination. Several readings of each extinction position should be taken.

10. When reading the extinction position of the surround, it is preferable that this position be as close as possible to that of the specimen. Moreover, the extinction positions of the object and its surround should be determined with no delay between the two steps.

CHAPTER 10

The Metallurgical Microscope

THE METALLURGICAL MICROSCOPE is an important instrument for the study and measurement of metals and such other opaque materials as minerals, ceramics, and textile fibers. By using the instrument with a suitable camera, permanent records can be made of the structure, grain size, adulterants, and other physical properties of metallographic specimens. In most cases, the specimens are opaque and can therefore be viewed only when illuminated by incident or reflected light, rather than transmitted light. Except for this difference in direction of the source of illumination, however, the procedures for using the metallurgical microscope are similar to those for the standard microscope.

OBTAINING, PREPARING, AND STORING METAL SPECIMENS

Cutting or Fracturing a Large Specimen

A small, representative metal specimen can be cut from a larger one with a hacksaw or an abrasive cutting machine. If the metal or alloy is brittle, small specimens can be obtained by fracturing the larger specimen with a hammer.

In cutting a metal specimen, care must be taken not to alter the original microstructure and characteristics or the metallographic properties. In most cases, high temperature is the major cause of such unwanted alterations. The microstructure, for example, may be grossly affected by heat. For this reason, the metal must be cut by a procedure known as *cold-working*, which involves cutting the metal slowly and directing a steady stream of cold water on the point of contact between the metal and the saw or abrasive wheel. The specimen must always be cut clean, without burrs.

SIZE OF THE SPECIMEN

The specimen should be neither too small nor too large, since small pieces are hard to handle in the grinding and polishing procedures described below and large ones require polishing of too great an area. A satisfactory size is usually from $\frac{1}{2}$ in. to 1 in. square by $\frac{1}{2}$ in. thick, with round or square contours. A specimen with dimensions in these proportions is of appropriate size to be adequate for the purpose and yet not require excessive polishing.

Grinding and Polishing the Specimen

The surface of a metal specimen must be prepared according to the procedure to be described in order to reveal the major characteristics of the metal. Accurate study of the specimen depends on the care with which this preparation is done.

The first step in preparing a metal specimen for examination is to grind it roughly with emery paper or sandpaper. Successively finer grades of paper are used until the metal surface is semipolished. A mechanical grinder is useful at this stage. Final polishing is then done with fine abrasives on cloths attached to the powered polishing wheel, which is called a *lap wheel*. This final polishing must produce a smooth mirror surface, free of scratches. Such a surface is essential to the last step, etching, which reveals the structural characteristics and flaws of the metal.

Great care must be taken in handling the specimen throughout the preparation. Since extraneous particles or grit may scratch the polished surface, the hands must be washed thoroughly after each step in the grinding and polishing procedure to remove the coarser abrasive particles so that they will not be mixed with the finer abrasive used in the next step.

HAND GRINDING

COARSE

The initial, coarse grinding of the specimen varies according to whether a motor-driven grinder is available or the grinding is to be done by hand. In the latter case, grind down the coarse surface irregularities on one face with a file. Do not exert excessive pressure during this step, since any new scratches will be difficult to remove. Moreover, excessive pressure may produce a deformed metallic surface layer that may give poor results when the polished surface is etched. Keep the specimen cool by frequent wetting during grinding. This rough grinding is continued until the surface is flat and free from major flaws. The specimen is then washed thoroughly in water before continuing.

SEMICOARSE

Carborundum or emery paper, in various degrees of fineness or grit, is used to remove any further, minor irregularities in the specimen surface. It is good practice to use a new sheet of abrasive paper for each specimen. The first sheet should be of no coarser than medium grade (No. 1 emery paper or No. 240 carborundum paper). Even Finer grades (No. 0, which is also called 1/0, emery paper or No. 320 carborundum paper) may also be used.

Lay the abrasive paper on a hard, flat, level surface, such as a sheet of plate glass or metal (see Fig. 130) with the abrasive side up. Hold

Fig. 130. Hand-grinding sheet.

the paper down with one hand and hold the specimen with the other. Place the specimen face down on the paper and, applying moderate pressure, draw it gently back and forth across the entire length of the paper so that it is ground along one direction.

Replace the first sheet with a finer sheet, such as No. 00 (2/0) emery paper or No. 400 carborundum paper, and again draw the specimen gently back and forth across the entire length of the paper, but at right angles to the original direction. The new, smaller scratches made by the finer paper will replace the larger ones made by the previous, coarser paper. Grind twice as long with each successively finer abrasive paper to remove previous scratches and any disturbed surface metal left by the coarser paper. Wash the specimen thoroughly in running water at each change of paper.

Soft metal samples or alloys may require lubrication or wetting of the abrasive paper for good results. This not only makes the grinding less fatiguing, but also cools the face of the specimen so that its structure is not altered or distorted. Lubricants for this purpose include water, liquid soaps, glycerin, low-viscosity petroleum oils, and kerosene.

FINE

Final, fine hand grinding is similar to semicoarse grinding, with increasingly finer abrasive papers being used, from extra-fine paper such as No. 000 (3/0) emery paper or No. 600 carborundum paper to No. 0000 (4/0) or 00000 (5/0) emery paper. As before, a coolant such as water should be used during the finer grinding to minimize any microstructural changes in the metal.

Fine grinding is continued until the surface of the sample reveals only very small, uniform scratches, with all coarser scratches removed, when examined with a magnifying glass.

The specimen is now ready to be polished. The final results achieved with polishing depend on the condition of the specimen after grinding, since the polishing operation cannot remove coarse scratches.

MACHINE GRINDING

Grinding by hand is adequate if only one or two specimens are to be prepared, but the use of an electric grinder is recommended. Such grinders have the abrasive paper affixed to the grinding wheel by means of a ring or other holding device. Some have two bakelite or metal wheels, with abrasive paper attached to both sides of each wheel (see Fig. 131).

Fig. 131. Electric grinder.

The motor is usually a two-speed type. The fast speed (about 600 rpm) is used for the coarse papers; the slower speed is used for the finer papers, since the latter tend to burnish the metal samples at high speeds. The specimen must not become overheated, nor should great pressure be used against the rotating wheel. The same precautions must be taken in machine grinding as in hand grinding.

POLISHING

An ideally prepared metallographic specimen must have a mirror-like surface, free of scratches, surface imperfections, and all traces of disturbed metal. Polishing produces such a surface by removing the fine scratches made during the final grinding. An electric polisher (see Fig. 132) is essential for this operation.

The polishing wheel is a disc that rotates in the horizontal plane and is covered with a polishing cloth, technically known as a *polishing lap*. Some polishers have variable-speed motors.

There are also automatic polishing units that have extension arms for polishing more than one specimen simultaneously. These arms hold the specimens over the polishing wheels with moderate

Fig. 132. Electric polisher.

pressure, automatically moving them back and forth across the wheels and rotating them to alter the polishing direction continuously.

POLISHING CLOTHS

Polishing cloths are of two kinds—shorter-napped, for rough polishing, and longer, softer-napped, for final polishing. A large variety of polishing cloths is available; the choice of cloth depends on the kind of metal specimen and the degree of polishing required. Most cloths do not require special treatment, but if a cloth is not sufficiently flexible or has hard spots, it can be soaked in boiling water to soften it.

Polishing cloths are sold under various trade names. "Selvyt" is a high-quality imported medium-nap cotton cloth for rough and final polishing. Nylon cloth, which has a napless, satin weave, is very durable and excellent for rough polishing. "Duracloth" and "Microcloth" are synthetic rayon cloths in which the fibers are bonded to a cotton-twill backing to form a long-napped cloth. They are used for final polishing. Billiard cloth is a high-quality wool fabric with a uniform medium nap, recommended for final polishing. Silk polishing cloths, without nap, are used for coarse polishing with abrasives. Canvas duck is a rough cotton cloth used primarily with various coarse grades of abrasives. "Kitten-Ear," a short-pile all-wool broadcloth, is recommended for final polishing operations.

USING POLISHING CLOTHS

When used for the first time, the flexible polishing cloth is soaked in water, stretched tightly over the polishing wheel, and secured firmly. A polishing abrasive is mixed with water to form a thin paste, which is applied with the fingers and worked uniformly into the cloth.

To preserve the cloth for future use, remove it from the wheel, wash it thoroughly in running water to remove any abrasive material and metal impurities, and store it moist until ready for use again.

ROUGH POLISHING

Polishing should always be done in a dust-free atmosphere, so that successive polishing laps are not contaminated.

The object of rough polishing is to remove the very fine scratches remaining from the last grinding operation. For this step, the polishing wheel is covered with a short-napped cloth, such as a broadcloth, billiard cloth, or lightweight canvas cloth. A fine abrasive powder, such as No. 600 silicon carbide powder, tripoli powder, alumina powder, or diamond dust, is mixed with water to make a thin paste. The paste is applied to the cloth and the wheel rotated at 400 to 500 rpm. The metal specimen is held against the rotating lap and moved continuously from one edge of the wheel to the center and back again. During this operation, more abrasive paste is added from time to time with a large pipette or other uniform dripping device. If the cloth is too wet, it will polish poorly and may cause pitting of the metal surface; if it is too dry, it will tarnish or dull the surface. The moistness of the cloth should be such that the surface of the metal specimen acquires during polishing a film of moisture that evaporates in 5 to 8 seconds after the specimen is removed from the wheel. The presence of sufficient abrasive in the cloth is indicated by a thin opaque film of abrasive on the metal surface.

Rough polishing takes about 10 minutes, after which the specimen is removed from the wheel, washed in running water, and examined for any remaining scratches with the aid of a good magnifying glass. The surface should be uniform, with a dull, mirrorlike appearance. It should be practically free of scratches from grinding, and only very fine scratches from the polishing abrasive are acceptable. If grinding scratches are still discernible, it would be best to repeat the last grinding operation carefully since additional polishing will not remove them. If the specimen is soft and the surface is likely to become pitted, the rough polishing stage is omitted.

FINAL POLISHING

This step removes any remaining small, fine scratches and produces a highly polished mirrorlike surface. A soft, long-napped polishing cloth is used.

Many fine abrasives are available for final polishing — among them iron oxide (rouge), magnesium oxide, chromic oxide, and alumina. The latter, a good all-purpose fine abrasive, is most often used for metallographic polishing of hard, medium, and soft metals. Polishing alumina can be purchased suspended in a slurry. Dry alumina is also available, and it is usually levigated to obtain the finer particles for polishing. This levigation is done by mixing the dry alumina with water and allowing the suspension to settle for a few minutes. The fine abrasive particles remain suspended, while the coarse particles sink to the bottom. The liquid with the suspended particles is poured off and used as the fine polishing abrasive. The remaining sediment can be levigated again to obtain suspensions of slightly coarser particles. Magnesium oxide, an abrasive with small particles, produces the finest finishes on soft, nonferrous metals or alloys. Since the particles are small, polishing may take relatively longer. Diamond abrasives, available in various particle sizes, are excellent, and they can be applied as a dust or, more economically, as a paste. Jewelers' rouge is also used for final polishing, but the procedure is slow and the rouge occasionally becomes embedded in the surface of the metal

It is recommended that all steps in final polishing be done as rapidly as possible, since prolonged polishing causes undesirable relief effects.

To polish a ground specimen, place the moistened cloth on the polishing wheel, clamping it on tightly. Rub the fine polishing abrasive into the fibers of the cloth. Then place the specimen on the rotating lap and polish it for about 5 minutes, moving it continuously from the edge of the wheel to the center and back to the edge. Toward the end of the final polishing, change the polishing direction occasionally by turning the specimen counter to the rotation of the lap.

Remove the specimen from the lap occasionally and inspect it with a magnifying glass to see whether the surface is free from scratches. If it is sufficiently polished, wash it thoroughly in warm running water, wipe the surface with a soft cotton swab moistened with alcohol to remove any adhering abrasives and foreign fragments, and, finally, rinse again in running water, moisten with a little alcohol, and dry immediately with a blast of warm air.

An adequately polished specimen will appear practically free from scratches when viewed under the microscope at 100X magnification. If scratches are still present, begin again with the rough polishing operation.

Certain precautions should be observed in handling polished specimens. Never touch the polished face with the bare fingers or with any implement that can scratch the surface. The fingers may leave a smudge, which will inhibit the chemical reaction in the etching operation.

ELECTROLYTIC POLISHING (ELECTROPOLISHING)

Electrolytic polishing is a process that removes a small amount of metal from the surface of the specimen, leaving a smooth, scratch-free surface with no deformation of the metal. This is accomplished by making the metal specimen the anode in an electrolytic cell containing appropriate electrolytes. When properly applied, electropolishing offers many advantages over mechanical methods. Mechanical polishing often deforms the metal surface so that it is not basically characteristic of the metal proper, whereas electropolishing produces no such deformation. The technique is easily learned and eliminates tedious work. Other advantages of the electropolishing technique are that many specimens can be polished at the same time and that electropolished metal surfaces rarely need repolishing to eliminate remaining scratches. The process is suited not only to such hard metals as stainless steels, but also to softer metals. The metal surface can be polished and etched in the same cell by decreasing the voltage to about a tenth of the voltage required for polishing and continuing electrolysis for a few more seconds.

There are, however, some disadvantages to electropolishing. Since many of the chemicals used as electrolytes are toxic and hazardous if not properly handled, a basic knowledge of chemistry is essential in using them. Also, a large variety of electrolytes may be needed for polishing different alloys and metals and special storage facilities are required. Electropolished specimens may show surface undulations rather than plane surfaces. These may be objectionable at low magnifications, but will not be at 100X magnification. Certain metals become chemically inactive when electropolished, making them difficult to etch. In some cases, electropolishing may produce a metallographic specimen in which the structure is not typical. Finally, the success of electropolishing depends on many critical variables, such as the proper kind and concentration of electrolyte, the correct solution temperature, current density, and voltage, and the time essential for proper electropolishing.

The suggested theory underlying the electropolishing technique is that it produces partial anodic decomposition of the specimen. The scratches can be said to consist of ridges and valleys, the former being dissolved away, the latter, between the ridges, being protected by the products of the chemical reactions.

Electropolishing Apparatus

The electropolishing apparatus (see Fig. 133) consists of two main parts—the electrolytic cell and a variable power source. The cell consists of a corrosion-resistant tank and a stirring device or impeller that directs a gentle flow of electrolyte against the sample, which is held by a fixed mask that exposes only the predetermined area to be polished.

The power supply may be batteries or a rectified alternating

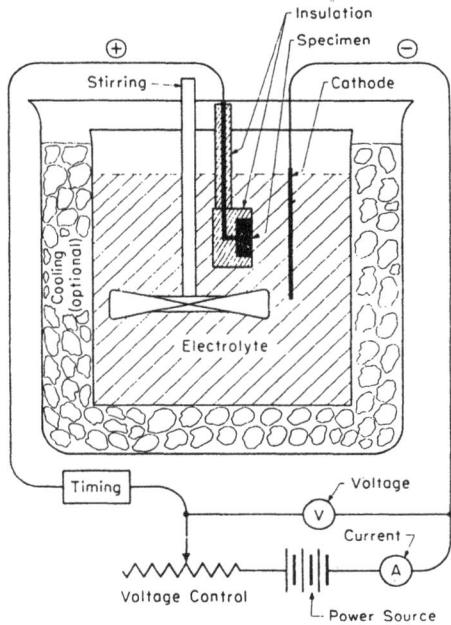

Fig. 133. Electropolishing
apparatus.

current, with an ammeter and voltmeter included in the circuit to control and indicate the voltage and current density. A timing switch to control the polishing cycle is generally included with the available power supply.

Procedure for Electropolishing

The instructions that follow are general and may differ slightly for different types of metal specimens.

The surface of the ground specimen must be thoroughly clean. The specimen is washed thoroughly in warm water to remove any remaining abrasive, then rinsed in acetone or carbon tetrachloride to remove any oily or greasy deposits.

The metal specimen is the anode. The choice of material for the cathode depends on the composition of the specimen, as does the choice of electrolyte. For example, a solution of chromic acid (620 g of chromic acid in 830 ml of water) can be used as the electrolyte for a stainless steel specimen. (Many other electrolytes can also be used for stainless steels—a mixture of phosphoric and sulfuric acids at several different concentrations, for example, or mixtures of sulfuric and hydrofluoric acids.)   An electrolyte commonly used for polishing both ferrous and nonferrous metals is a mixture of absolute methanol (660 ml) and nitric acid (330 ml).

If the voltage is too low, the surface of the metal may be etched and give a rough or matte appearance.   If it is too high, excessive gas bubbles will appear at the anode and cause uneven polishing or pitting of the specimen.

When polishing is complete, turn off the current and remove the specimen.   Wash the metal in a stream of warm water, then dry in a blast of warm air.   Store the specimen as described later in this chapter.

MOUNTING SMALL SPECIMENS

Small specimens that are too small to be held conveniently during the grinding and polishing operations—thin sections, for example — must be mounted and held rigidly in some device.   A few suggested methods for mounting specimens are described below.

FUSIBLE AND CAST MOUNTS

Low-melting alloys or mounting media that fuse easily at a moderate temperature and then solidify back to their original state as the temperature is lowered are used for this type of mounting. Low-melting alloys containing bismuth, such as Wood's metal or Lipowitz alloy, melt at about 70°C and can be easily manipulated. Solder (50% lead, 50% tin) is an alloy that is used frequently to mount ferrous metals and ferrous alloys.   Other materials such as sealing wax, DeKhotinsky cement (hard and medium grades), plaster of Paris, dental cement, and litharge-glycerin cement are also extensively used for mounting specimens.

In choosing the fusible material, consideration must be given to the extent to which the material will resist attack by electrolytic or etching chemicals.   The specimen must be properly mounted so that the mounting material does not come into contact with the abrasive papers during the grinding operations, since these materials may clog the abrasive papers.

These resin materials are used in the same manner as cast and cement materials. *Thermosetting plastics*, which are permanently set by heat, are generally composed of phenol-formaldehyde or aniline-formaldehyde resins. Because a mount made of thermosetting resin cannot be reshaped once it has been fully solidified and set, it can be removed from the mold as soon as it sets.

Thermoplastics are synthetic resins that are softened by heat and become rigid at ordinary temperatures. Since thermoplastics are affected by heat, a mount made of this material is usually cooled to 75°C or less before it is removed from the mold. Polystyrene, Tenite (a cellulose-base material), and lucite (a methyl methacrylate resin) are some thermoplastic resins that can be used for mounting small specimens.

A special mounting press is essential when mounting specimens in a thermoplastic resin. Such presses apply heat and pressure simultaneously to the resin. Thermoplastic resins are ideally suited for mounting, since they cool to a rigid, hard mass, do not clog the emery paper, and are not extensively attacked by most etching chemicals.

Liquid resins are also available. They are poured over specimens in molds and then heated. This treatment hardens the resin around the specimen. An advantage of such resin is that no press is required and production time is shorter.

MECHANICAL MOUNTS

Small specimens can be advantageously mounted in a laboratory screw-clamp or between filler strips that are of approximately the same hardness as the specimen and are held together as tightly as possible by bolts or other devices.

Etching the Specimen

Microscopic examination of a polished specimen that has not been etched reveals very little structural detail because a thin, amorphous metallic layer is formed on the metal surface during grinding and polishing. This layer is sufficient to conceal almost completely the crystalline structure and must be removed if the microstructure of the manufactured metal is to be revealed.   This is done by attacking the polished surface with chemicals called *etching reagents*, many of which are selective in their attack on the various constituents of the fabricated metal and thus produce a crystalline pattern on the metal surface.   Under the microscope, the effects of the etching reagent on the metal structure can be seen as differential orientation of the grains, and the grain boundaries are made apparent by dark, narrow lines.   These effects provide better optical contrast between adjacent structures.

Etching reagents are dilute solutions of alkalies, organic and inorganic acids, and other chemicals. Liquids used as solvents include water, alcohol, glycerin, or a mixture thereof.   Alcohol solutions are extensively used since alcohol is a good wetting agent and will cut oil or grease films to permit uniform etching of the metal surface.

Different types of metals or alloys require different etching reagents to reveal their crystalline structure. Etching reagents are also selected to bring out certain desired characteristics of the metal structure. For example, a solution of chromic acid is excellent for etching copper and its alloys, but is unsuited for aluminum and its alloys or for iron and steels. Many different etching formulas can be used for a given metal, but the one chosen should reveal the desired crystalline characteristics.

Once the correct solution has been chosen for a specific job, the directions for its use must be scrupulously followed.   Before etching the polished specimen, clean it thoroughly with acetone or carbon tetrachloride to remove any grease or oil deposits and thus ensure uniform etching.   Rinse the specimen in running water and wipe it with a swab of wet cotton.   Then rinse it in alcohol and dry it in a blast of warm air.

To ensure optimum results, the concentration and temperature of the etching reagent, and especially the time allotted for etching, must be rigidly controlled.   The time should be sufficiently long to give a sparkling etched surface, with good contrast of the various constituents.   The time also depends on the magnification under which the specimen is to be viewed.   The higher the magnification, the less time necessary for etching, since high magnification does not require deep etching. If in doubt as to the time necessary, it is better to underetch rather than overetch, since overetching may partially obliterate the crystalline structure of the specimen.

The etching reagent can be applied to the polished surface either by *swabbing* or by *immersion*. Swabbing generally gives more uniform results and is used when photomicrographs are to be taken or some of the constituents of the metal are to be identified. Immersion is used for microscopic examination, in which low magnifications (generally 10X or less) are used. This method is convenient for routine examination, which is sufficient to give the desired information. These two methods are described below.

SWABBING

Wrap a tuft of absorbent cotton around a wood applicator, dip the swab into the etching reagent, and swab the metallographic specimen surface for the specified length of time. Dip the swab into the etching solution frequently, so as to keep the cotton saturated and bring fresh reagent to the polished metal surface.

When etching is judged to be complete, rinse the specimen immediately in running water to stop any further chemical action. Rinse again in alcochol and dry in warm air. If microscopic inspection shows that the specimen is not sufficiently etched, repeat the procedure. If the etching reagent selected does not produce the expected results, repolish the specimen and try another etching reagent.

IMMERSION

Using a pair of metal tongs, place the specimen face up in a small glass vessel containing sufficient etching reagent to cover the specimen. Do not hold this specimen with the tongs during immersion, since the metal of the tongs may affect the etching process. Move the specimen continuously in the etching solution, so that any gas bubbles evolved will not cling to the surface and cause uneven etching.

As the reagent acts on the metal, the surface becomes dulled, and the degree of dullness is a rough guide as to whether the specimen is sufficiently etched. After etching, stop the chemical action, rinse and dry the specimen, and inspect it in the same manner as described for the swabbing method, repeating the etching procedure or repolishing the specimen and trying another etching reagent if necessary.

Storing Metal Specimens

Specimens are usually stored either etched or unetched in a chemical desiccator until ready for use. The desiccants used are "Drierite," anyhydrous calcium, or magnesium chloride. Etched specimens can often be more easily preserved by dipping the metal surface into a thin, clear metal lacquer, draining for a few seconds, and allowing to dry. The lacquer can also be painted on the surface with a soft brush. If lacquer is not available, a thin smear of petrolatum, oil, or soap solution will prevent corrosion.

METALLURGICAL MICROSCOPY

Metallurgical Microscopes

The standard compound microscope may be readily converted into a metallurgical microscope (see Fig. 134). To do so, a vertical illuminator is screwed into an opening of the microscope tube just above the objective lens. The substage condenser is removed, since the vertical illuminator is the source of illumination. A movable stage rather than the standard movable eye tube is a necessity for a good metallurgical microscope, since any movement of the objective when bringing the specimen into focus may impair the best adjustment of the vertical illuminator in the microscope. Focusing of the specimen is accomplished by moving the stage by means of the substage coarse and fine adjustments.

The *inverted metallurgical microscope* (see Fig. 135) is the type used most extensively for metallurgical work. This instrument, together with the vertical illuminator, light source, and camera attachment for taking photomicrographs, is also called a *metallograph.*

The inverted microscope differs from the standard type in that the objective is held rigidly in a vertical position under, rather than above, the movable stage. Specimens of any size can easily be placed face downward on the stage, perpendicular to the optical axis, with no

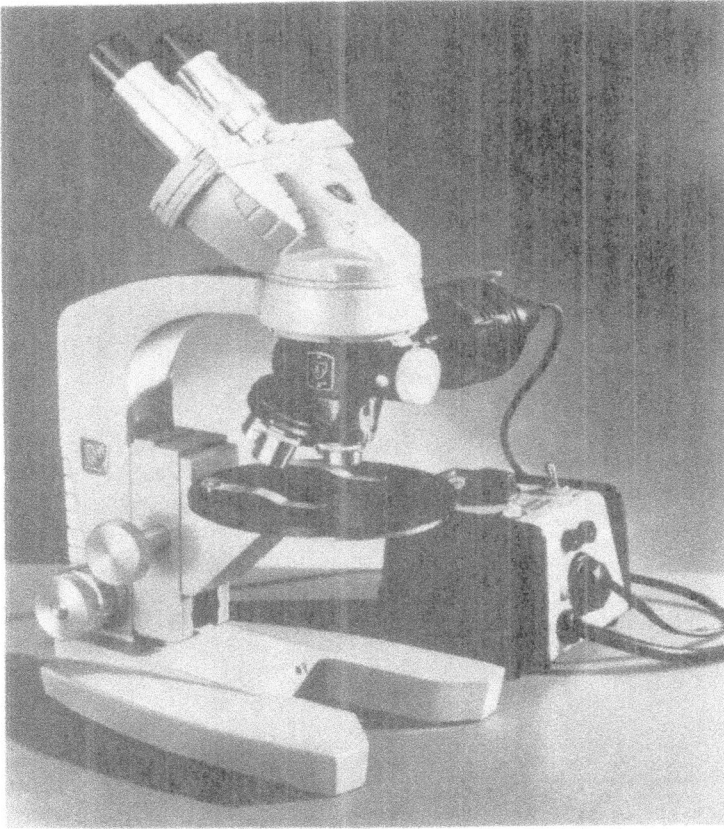

Fig. 134.   Metallurgical microscope and vertical illuminator.

problem of adjustment.   The stage can be moved up or down by means of the coarse and fine adjustments to bring any part of the specimen into focus.   The instrument permits permanent alignment of the microscope, vertical illuminator, source of illumination, and camera attchment, assuring brilliance and sharpness of detail in photomicrographs.   Its versatility permits the use of polarized light, phase-contrast, and dark-field illumination.

1) Stage
2) Circular stage plate
3) Mechanical stage control
4) Coarse focusing control
5) Fine focusing control
6) Revolving nosepiece and objectives
7) Bulb socket holder
8) Illuminator and condenser unit
9) Bulb centering screw
10) Field iris diaphragm
11) Aperture iris diaphragm

12) Filter holder
13) Path selector slideway
14) Plane glass reflector lever
15) Slot for Polaroid analyzer
16) Knurled ring to attach binoculars
17) Microswitch and cable release
18) Binocular body
19) Interpupillary distance adjustment collar
20) Diopter adjustment collar
21) Camera coupling tube
22) 35mm camera back

Fig. 135. Inverted metallurgical microscope (metallograph).

Metallurgical Objectives

Although the optical elements of the objectives commonly used in metallography are basically similar to those in standard microscopes, they differ in design because of their specialized use.   Since they are generally used with a vertical illuminator, they are built with a short mount so that the back lens of the objective is closer to the vertical illuminator. Metallurgical objectives are also designed for a specific tube length, which may vary with different makes of microscopes. They are corrected for operation without cover glasses or slides, which are not used with metallurgical specimens because the intense source of light would cause internal reflections to bounce off the glass surfaces, thus impairing the image and resolution of the metal surface.

Because of the design of these objectives, the revolving nosepiece is generally not used with the metallurgical microscope. Instead, each objective is placed in the optical path by snapping it into a receptacle attached to the vertical illuminator.

Metallurgical Eyepieces

Metallurgical microscopes are either monocular or binocular. The conventional eyepiece is satisfactory for direct visual examination of opaque objects, but different types of oculars are used for photomicrography. Compensating eyepieces, chromatically over-corrected, are used to compensate for color aberrations inherent in apochromatic objectives. Amplifier eyepieces are also used extensively to produce a "flat field," so that the image is in focus not only in the center of the field, but also at its edges. The amplifier eyepiece is used only when photomicrographs of metallurgicla specimens are to be taken.

Methods of Illumination

The methods of illumination used with the metallurgical microscope become especially important when a fine photomicrograph of an opaque specimen is to be made. Severel methods are discussed in the following paragraphs.

VERTICAL ILLUMINATION

The opaque specimen can be illuminated by incident light from above and from different angles, thereby achieving various effects. Vertical illumination is used most extensively with metallurgical specimens, however, since it shows the specimen under flat lighting, without shadows, and is therefore an ideal method of revealing metallurgical characteristics.

To obtain vertical illumination, a beam of light must be directed vertically along the optical axis of the microscope, through the objective, and on to the surface of the opaque specimen. The commercial vertical illuminator is therefore positioned in (and in most cases built into) the microscope tube. The source of illumination used with the vertical illuminator may consist of a filament bulb, a diffusion disc to diffuse the light, and a green-yellow light filter. Since the intensity of light from this source is low, it is suitable only for direct visual work. For photomicrography, a more intense source is necessary, and in most cases an additional light source is used.

BRIGHT-FIELD VERTICAL ILLUMINATION

Bright-field vertical illumination is most often used for metallurgical specimens. The image appears dark against a lighter background (see Fig. 136). The path of incident light with the vertical

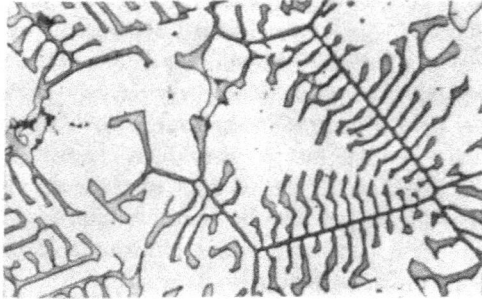

Fig. 136. Photomicrograph of an aluminum specimen under bright-field vertical illumination (250X).

illuminator is as follows: A reflector, which is either a clear plane-glass inclined at a 45° angle or a prism, is situated behind the objective in the noninverted metallurgical microscope. The beam of light that strikes the reflector is reflected at right angles to its original direction, down through the microscope tube, and through the objective to strike the specimen. This beam of incident light illuminates the specimen and is simultaneously reflected upward through the objective and again passes through the microscope tube to the eyepiece.

In this system of illumination, the objective serves two purposes. It condenses the incident light on the specimen to illuminate it and then magnifies the image, as in the standard microscope. Under these circumstances, the incident light passes through the objective twice.

Prism and Plane-Glass Reflectors

The *prism reflector* (see Fig. 137) is mounted in the vertical illuminator so that half the area of the objective is covered. The incident

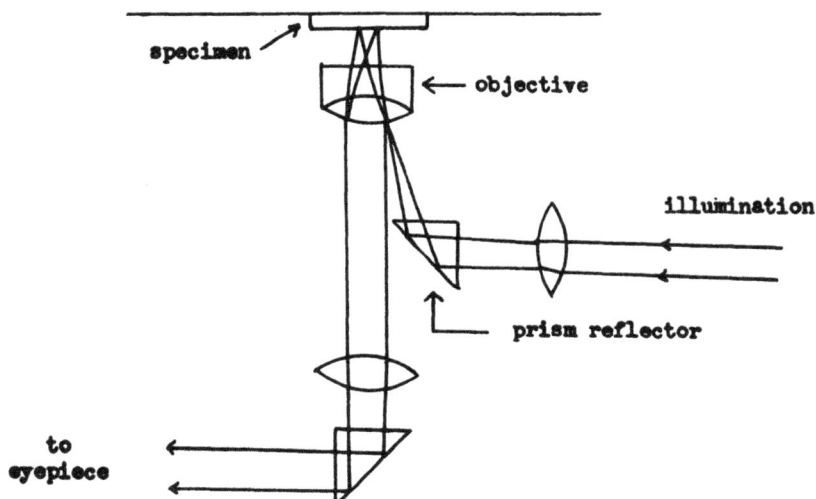

Fig. 137. Diagrammatic representation of vertical illumination with the prism reflector.

light is reflected through this half, while the other, unobstructed half allows reflected rays from the surface of the specimen to pass through to the eyepiece.

Since the prism reflector decreases the optimum efficiency of the objective by 50%, the numerical aperture is decreased proportionately, with a loss of resolution in the specimen. This loss is not too significant under low magnifications but image clarity and definition are impaired under higher magnifications. As a rule, the prism reflector is not used in photomicrography.

At lower magnifications the prism reflector gives more brilliant images, with better contrast, than the plane-glass reflector. The prism, however, must be properly aligned so that the specimen is evenly and uniformly illuminated. This is done by rotating the shaft that supports the prism through a small angle.

The *plane glass reflector* (see Fig. 138) is more advantageous for higher magnifications, since it does not decrease the numerical aperture of the objective and thus gives greater resolution.

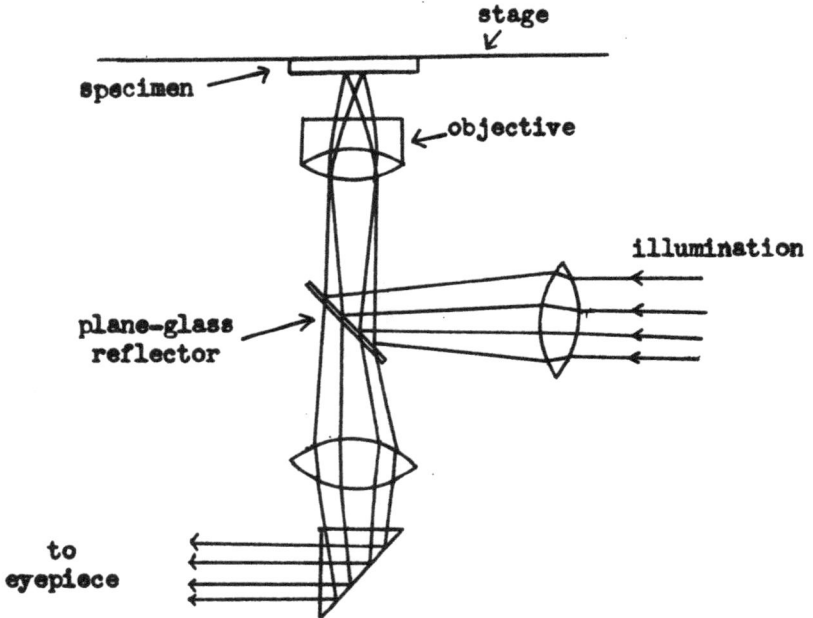

Fig. 138. Diagrammatic representation of vertical illumination with the plane-glass reflector.

Both the prism and the plane-glass reflector must be aligned so that they are inclined 45° from the incident beam. This alignment is accomplished by noting the position of a point of light transmitted through a pinhole source; the point should be positioned at the center of the field of view (see Fig. 139).

Since the plane-glass reflector reflects only about 20% of the illumination to the surface of the specimen because of the multiple reflection of the incident light from the glass surface, this type of reflector is not so efficient as the prism. Consequently, a more intense light source may be necessary for good observation of the specimen.

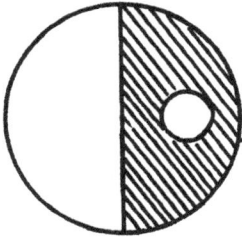

Correct (Prism) Incorrect (Prism)

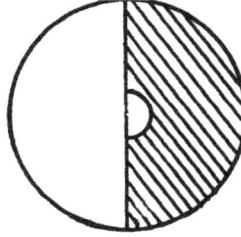

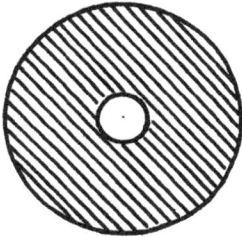

Correct (Plane-glass) Incorrect (Plane-glass)

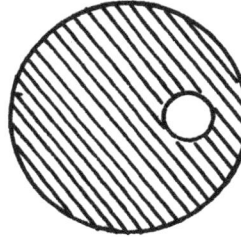

Fig. 139. Diagrammatic representations of the adjustment of the prism (top) and the plane-glass reflector (bottom) by means of a point of light.

Ultropak Vertical Illuminator (Leitz)

Another type of vertical illuminator is the *Ultropak*, which is manufactured by E. Leitz and Co. It is designed to illuminate the opaque specimen with rays of incident light and simultaneously keep the light from falling directly on the front lens of the objective. Incident illumination, accomplished by means of ring condensers surrounding the objective that reflect the light downward directly onto the object, is used only for image formation (see Fig. 140).

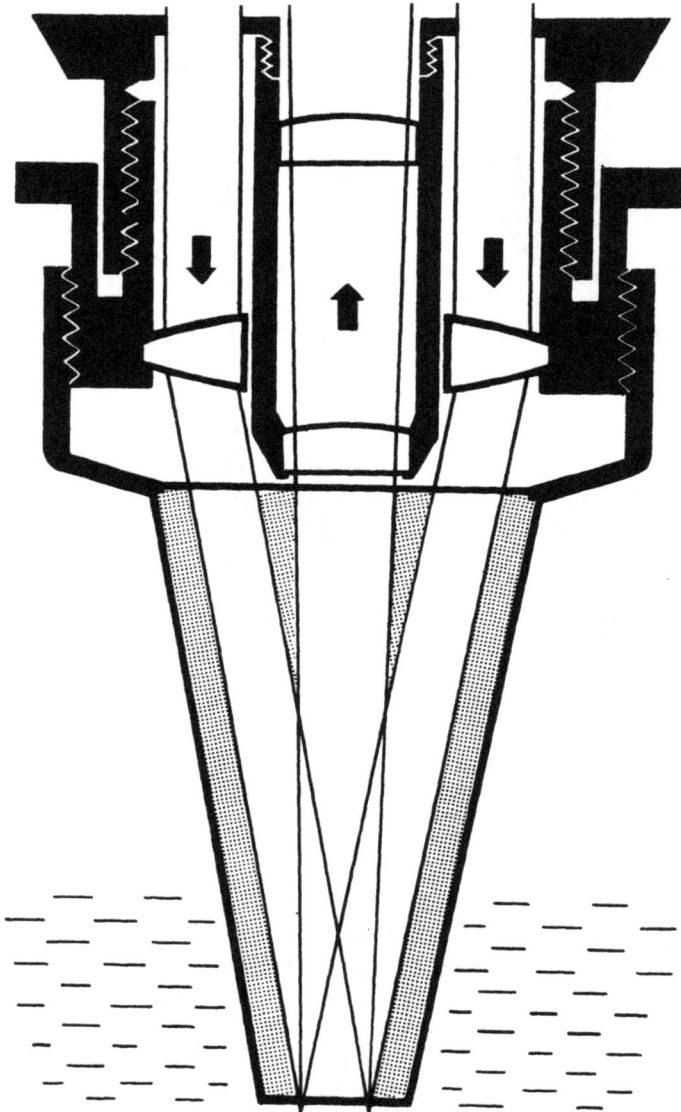

Fig. 140.   Cross section of the Leitz Ultropak vertical illuminator.

DARK-FIELD VERTICAL ILLUMINATION

Special apparatus is necessary to produce dark-field illumination in the metallurgical microscope. In bright-field vertical illumination, the objective serves not only as a condenser for focusing the light on the surface of the specimen, but also as an objective lens for forming and magnifying the image.

In dark-field vertical illumination, the objective does not serve as a condenser but is bypassed as the light travels down the microscope tube. The reflected light rays pass through a polished cemented prism and are reflected from its surface first to a special convex spherical mirror and then to a concave spherical mirror, which finally brings the incident light to the specimen. The rays are thereby deflected onto the specimen at an oblique angle, so that none of the light that is specularly reflected from the specimen passes through the objective.

Since dark-field vertical illumination uses the entire aperture of the objective to magnify the image, the numerical aperture is greater than with bright-field vertical illumination and will thus produce better resolution of the specimen image. Such details of the etched surface as grain boundaries, lines, and deformations appear bright, while the polished surface is dark and not visible (see Figs. 141 and 142). With bright-field illumination, the reverse is true.

ILLUMINATION WITH POLARIZED LIGHT

Polarized light can be easily obtained with the vertical illuminator by inserting a polarizing device into the filter holder, which is positioned between the light source and the vertical reflector. Some metallurgical microscopes have a built-in permanently aligned polarizer and analyzer equipped with a graduated 360° rotating mount.

In use, the polarizer is rotated while the specimen is under observation to determine the position at which the specimen appears brightest. At this position, the plane of vibration of the polarized light is exactly perpendicular to the plane of incidence of the light from the vertical reflector. As the etched specimen is rotated in the polarized light, different contrasting effects of the grain boundaries are seen and the actual grain inclusions of the specimen are more noticeable. Glare is also eliminated by the polarized light.

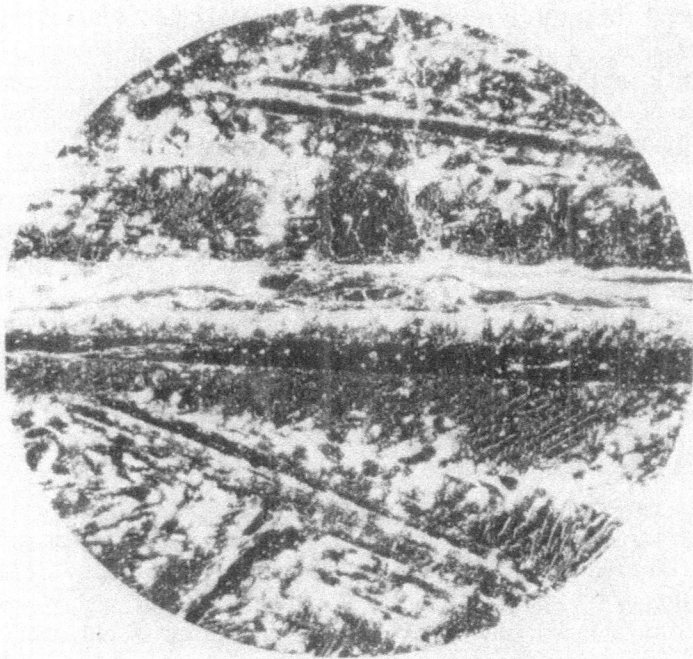

Fig. 141. Photomicrograph of a cast iron specimen under dark-field vertical illumination.

Anisotropic opaque minerals can be studied under vertical illumination and polarized light. Full-wavelength and quarter-wavelength polarizing compensators, which effectively produce slight differences in birefringence, are used in the same manner as described for transparent minerals in Chapter 7.

Polarized light can be employed to differentiate, to identify, and to determine the opaque characteristics of uniaxial and biaxial opaque minerals.

PHASE-CONTRAST ILLUMINATION

Phase contrast can also be used with the vertical illuminator to obtain greater contrast, especially for producing finer photomicrographs. To obtain phase contrast, an annular phase diaphragm centered in a special mount is inserted in the vertical illuminator and a

Fig. 142. Photomicrograph of a stainless steel specimen under dark-field vertical illumination.

corresponding phase plate is moved into the optical path.   As in the phase-contrast microscope, a telescopic eyepiece is used to check the concentricity and alignment of these two optical elements. The annular diaphragm is sharply focused on the phase plate by means of a special focusable lens positioned in the vertical illuminator.

OBLIQUE ILLUMINATION

Oblique illumination is used to accentuate certain features or details of an opaque specimen by casting additional shadows over it. Oblique illumination is also advantageous in that it achieves a slight increase in contrast and reveals details that are not revealed under normal illumination conditions.

The general method of obtaining oblique illumination is to decenter the aperture diaphragm of the vertical illuminator. The extent of such decentering determines the obliquity of the rays striking the specimen and the microscopic effects that are achieved.

Using the Metallurgical Microscope

The specimen is mounted on the movable stage of the microscope, which can be racked up or down so that the opaque specimen can be easily brought into focus. The correct short-mount, achromatic objective is inserted in the change receptacle provided in the metallurgical microscope for mounting the objective.

The surface of the specimen must remain perpendicular to the optical axis of the microscope, so that the specimen remains in focus when it in moved laterally in any direction and need not be refocused. The problem is negligible with the inverted microscope, since the etched surface of the specimen is flat and is placed face down on the stage. With the standard metallurgical microscope, however, the unpolished, unetched surface of the specimen is placed on the microscope stage and this surface is not always flat.

A simple method of solving this problem is that of mounting the specimen in a plastic material (such as modeling clay). To do this, place a ring, of a circumference slightly larger than the specimen and with edges exactly even, on a flat piece of plate glass. Partially fill the ring with the plastic material, and then insert the specimen in the following manner: Place the specimen face up in the center of the clay. Put a piece of soft facial tissue on the etched surface and lay a flat piece of glass on this tissue. Gently apply pressure on the upper glass until the etched surface is parallel to the upper edge of the ring. Remove the tissue and glass.

Rotate the coarse adjustment knobs until the specimen is brought to approximate focus; then bring the specimen into fine focus with the fine adjustment.

The light source is transmitted through the vertical illuminator so that the specimen is adequately illuminated. If the illumination is too intense, place a neutral-tint filter between the light source and the reflector of the vertical illuminator to achieve uniform illumination. It may be necessary to rotate the reflector slightly to achieve optimum illumination. Illumination can be more readily controlled with the attached vertical illuminator than with the standard microscope illuminator used for transmitted light.

As noted at the beginning of this chapter, the operation of the metallurgical microscope is similar to that of the standard compound microscope, except that incident light is used for viewing opaque objects. In most cases, a photomicrograph of the etched surface is prepared as a permanent record of the specimen. Photomicrography is discussed in the next chapter.

CHAPTER 11

Photomicrography

PHOTOMICROGRAPHY IS THE TECHNIQUE of photographing the magnified image of a specimen through the microscope. Its primary use is in making permanent records of fragile or perishable specimens. Photomicrographs permit more thorough study and are especially valuable in the comparative study of different specimens. (It should perhaps be noted here that the terms *microphotograph* and *microphotography* are sometimes used interchangeably with *photomicrography* and *photomicrograph*. The first two are technically imprecise, however, and should therefore not be used in this context.) They are also useful as illustrations for works on microscopy. A representative sampling of photomicrographs is presented in Figs. 143–149.

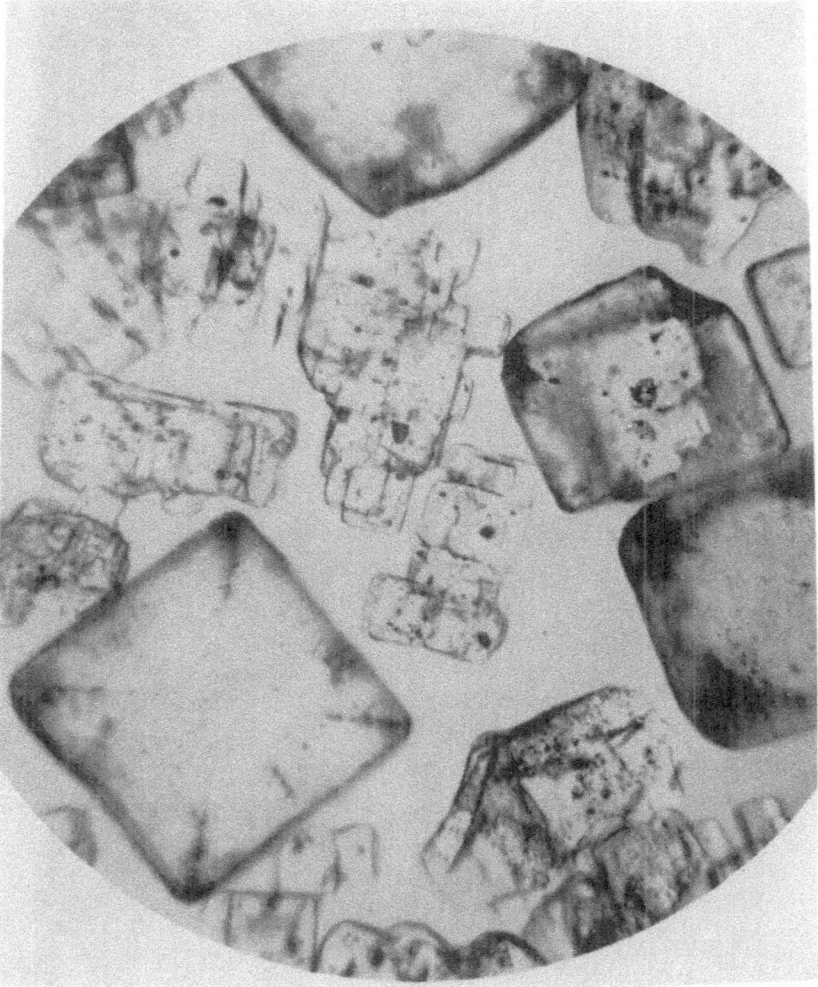

Fig. 143. Crystals of sodium chloride (table salt).

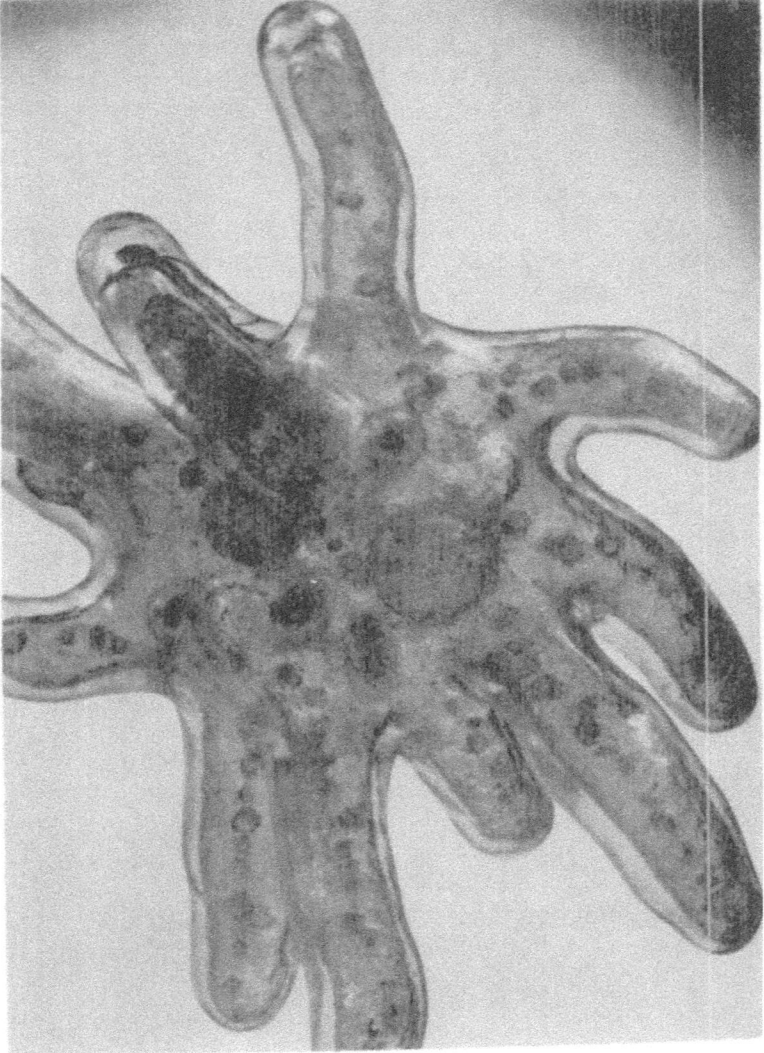

Fig. 144.   Amoeba.

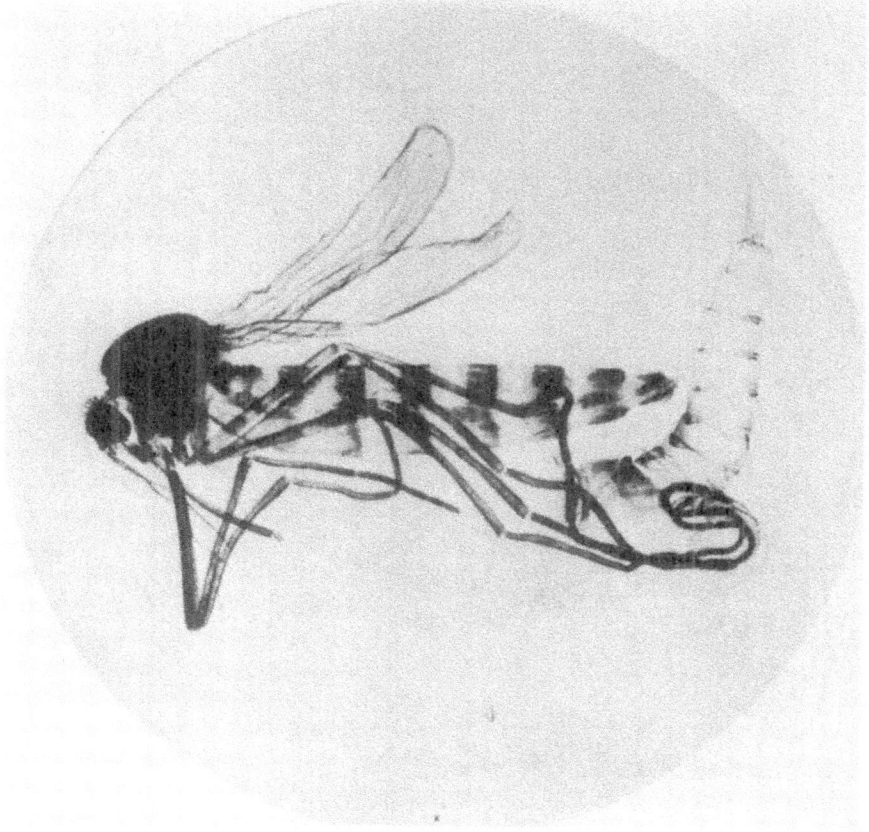

Fig. 145. Mosquito.

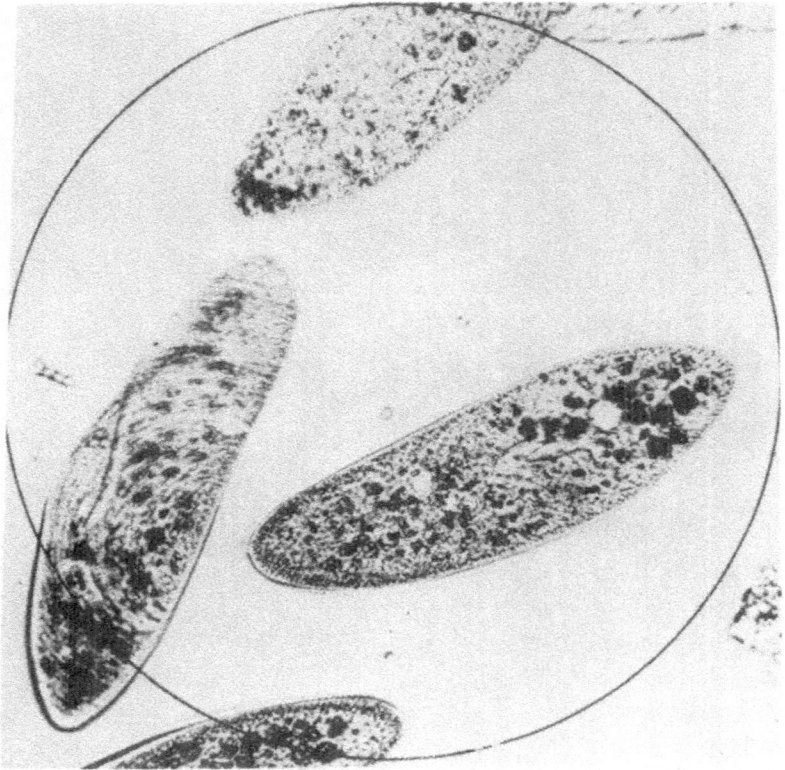

Fig. 146.   Paramecia.

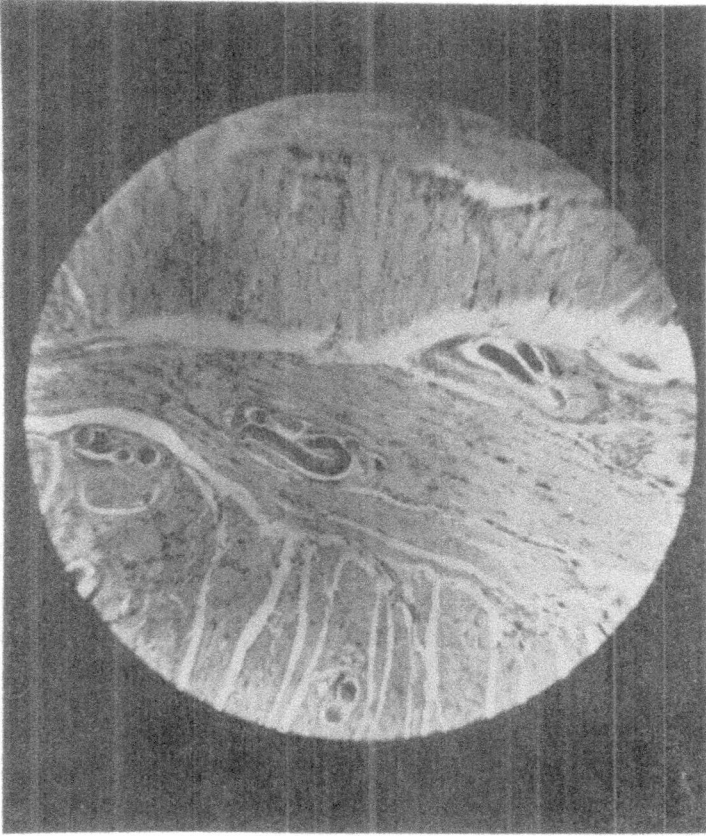

Fig. 147.   Trichinella *larvae.*

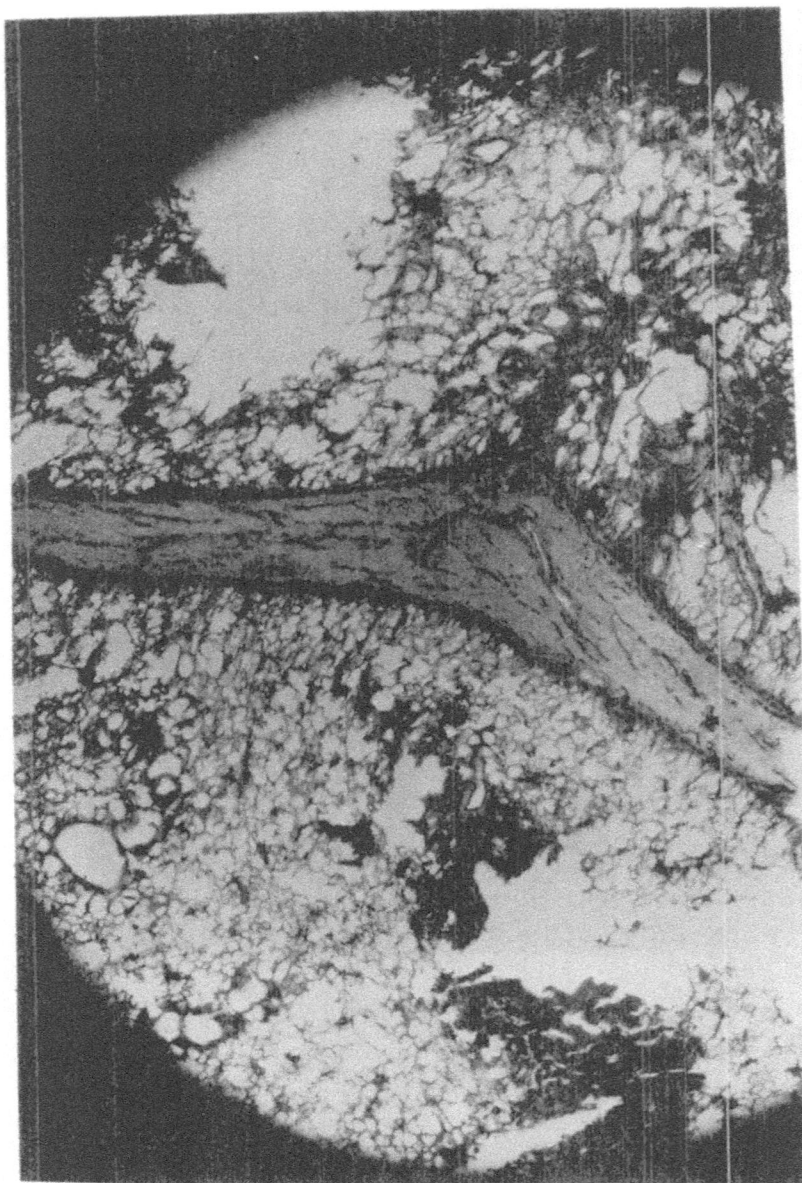

Fig. 148. Animal lung section.

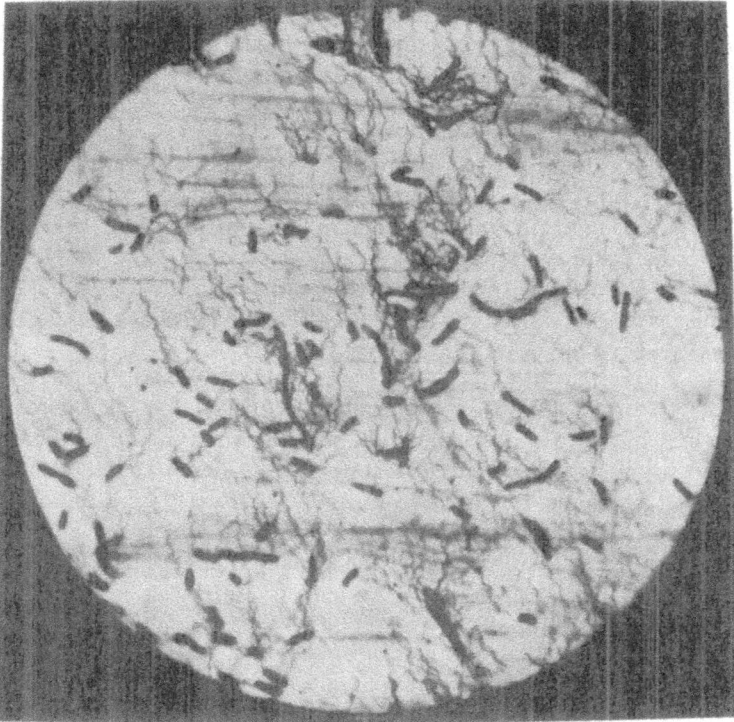

Fig. 149.   Typhoid bacilli.

SELECTING A CAMERA

It is possible to take photomicrographs with an ordinary camera attached to the body tube of a standard compound microscope. A simple setup using this method lets the microscope eyepiece act as the camera lens, while the fixed lens of the camera, which is mounted over the microscope body tube, projects the specimen image through the camera and onto the film.   There are, however, specially designed photomicrographic cameras that can be more easily adjusted.   The selection of the camera to be used is governed mainly by cost and the use for which the photomicrographs are intended.

Ordinary (Eyepiece-Attachment) Cameras

Any available camera, fixed- or variable-focus, can be used for photomicrography. A camera with a nondemountable lens can be used without removing the lens.

Whatever camera is used, it must be held rigidly in the correct position over the microscope eyepiece by some means, so that there is a light-tight connection between camera and microscope. A common laboratory stand will support the camera setup adequately, but a microscope adapter with a camera collar for mounting the camera over the microscope body tube is more satisfactory. With such an adapter, the microscope supports the camera.

For photomicrography, the specimen is placed on the microscope stage and brought into sharp focus. Since the visual image of the specimen as seen through the microscope is considered to be at infinity, the camera must also be focused at infinity before it is placed over the microscope. Also, the lens diaphragm must be set at the "wide open" position so that the full image of the specimen falls on the film.

Light rays emerging from the eyepiece come to a focus at a certain distance above it called the *eyepoint* (see Fig. 150), and the camera must be positioned over the eyepiece at this distance. The eyepoint can be determined by placing a ground-glass plate over the eyepiece and moving the plate up and down until the image is sharply focused on it. Since the specimen image diverges and becomes magnified above this point, the camera lens should be positioned at or near the eyepoint to secure a sharp, good-quality photomicrograph of the entire specimen. If the lens is placed too far above the eyepiece, the photomicrograph will not be sharp and a portion of the specimen image will be cut out.

Microscope companies make provision for attaching various types of ordinary cameras — such as a 35-mm camera, a Polaroid camera, or even a fixed-bellows camera, fitted with a cut-film holder and a ground-glass back to aid in focusing the image—directly to their instruments, with some means of focusing and viewing the specimen. Since the film plane in these types of cameras is fixed and cannot be adjusted, they are called *eyepiece-attachment cameras* when used in this way for photomicrography.

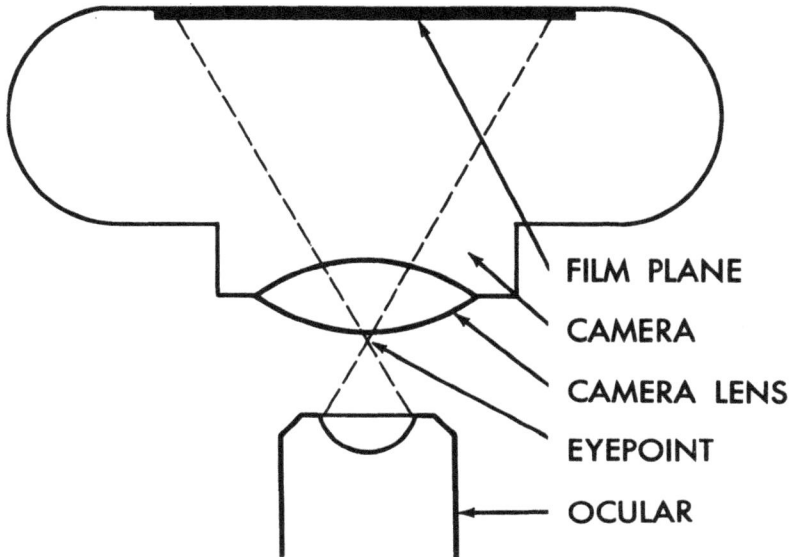

Fig. 150.   Diagrammatic representation of the eyepoint of the microscope.

Such a camera is attached to a special viewing head with a collar that fits directly over the microscope eyepiece (see Fig. 151). The viewing head incorporates a beam-splitting device, which transmits about 90% of the light to the camera and the remaining 10% to a focusing screen that can be viewed with a magnifier lens attached to the viewing head (see Fig. 152). The specimen image can be easily brought into focus, since it is continually visible on the glass screen before and during the photographic procedure and is not affected by the operation of the shutter release.

Eyepiece-attachment cameras, particularly those that use roll film (for example, 35-mm and Polaroid cameras), are especially suited for photographing live microorganisms, since they can take sequential photomicrographs easily. Roll-film cameras are versatile. They are also excellent for taking colored photomicrographs, since the film is relatively inexpensive and the resulting image is of high quality and can therefore be easily enlarged. They are convenient to use with most microscopes and give photomicrographs of nearly professional quality.

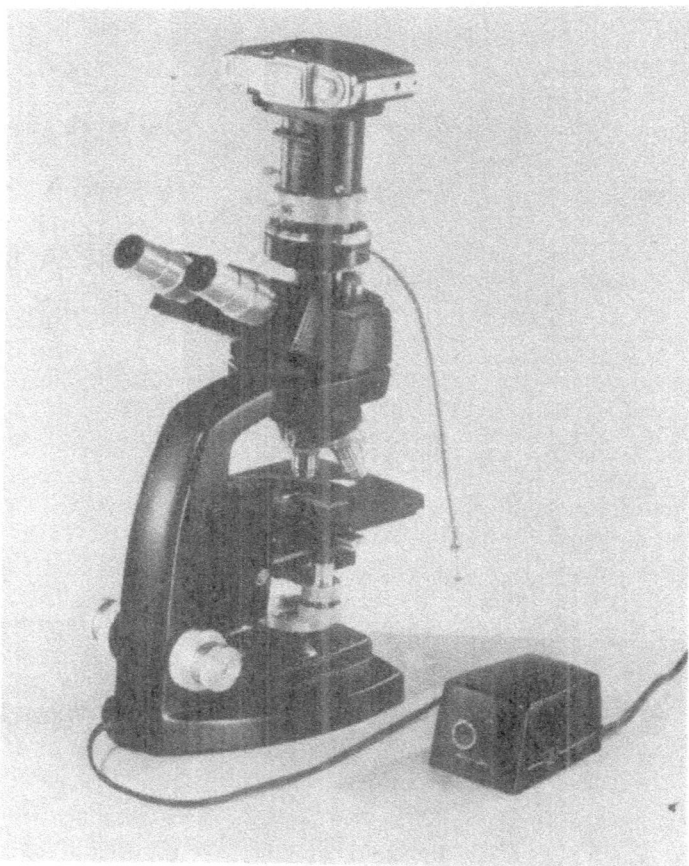

Fig. 151. Microscope with a 35-mm camera mounted on a viewing head.

There are limitations to the use of eyepiece-attachment cameras, however, the major one being that they cannot be adjusted for

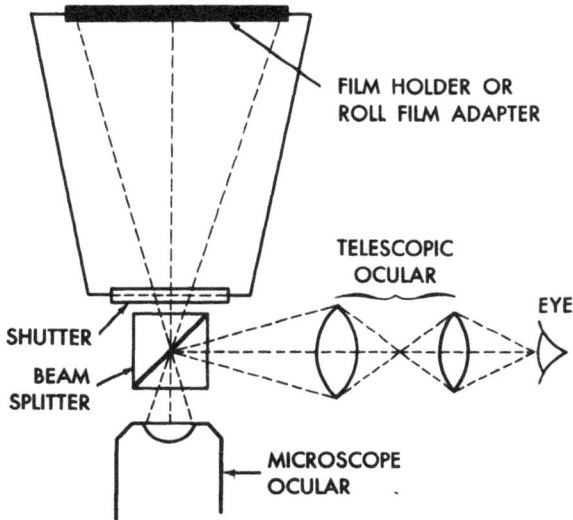

Fig. 152. Diagrammatic representation of an eyepiece-attachment camera with a beam-splitter.

magnification. In most cases, the magnification of the photographed image is significantly less than that provided by the microscope, which proves a drawback when very high magnification is desired. Use of a 10X objective and a 10X eyepiece, for example, will in theory produce an image 100X in diameter approximately 10 in. from the eyepoint. The focal length (the distance from camera lens to film) of nearly all such cameras is much less than 10 in., however, with the result that the photographed image is considerably smaller. Assuming a focal length of 38 mm (1.5 in.), the magnification of the specimen image on the film will be (1.5/10) × 100, or 15X—not the original 100X. The only way of increasing the magnification is to use a higher-power objective, or a higher-power eyepiece, or both. Since the negatives produced by these cameras are usually small, this reduction in magnification means, in many cases, that the photomicrograph obtained must be greatly enlarged, and a loss of resolution and definition may result.

Another limitation is that these cameras cannot be adjusted to photograph only the desired portion of the specimen image. Generally, almost the entire field of view must be photographed.

Finally, if the camera is used with its own lens, internal reflections within this lens may degrade the photomicrograph.

Photomicrographic Cameras

VERTICAL-BELLOWS CAMERA

The photomicrographic camera with an adjustable bellows and with no lens (see Fig. 153) is preferred for professional photomicrography,

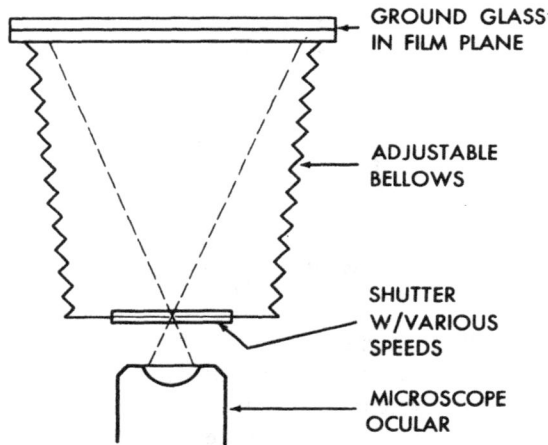

GROUND GLASS IN FILM PLANE

ADJUSTABLE BELLOWS

SHUTTER W/VARIOUS SPEEDS

MICROSCOPE OCULAR

Fig. 153. Diagrammatic representation of an adjustable-bellows photomicrographic camera.

since it is far more versatile. This type of camera permits varied ranges of magnification, selection of objectives, and adjustment of the portion of the specimen to be photographed. The camera is supplied with a ground glass for focusing and composing the image and incorporates a variable-speed shutter for controlling exposure time. The bellows camera accommodates 4-by-5-in. film or larger, these larger film sizes meaning that greater definition results if the photomicrograph is enlarged.

The bellows is provided with a plain back, consisting essentially of a ground glass focusing screen that can be removed and easily replaced by a single or double film-plate holder. Some cameras can be fitted with a reflex back, which reflects the image by means of a mirror, thus permitting the worker to remain seated while using the camera.

To eliminate vibration, the vertical-bellows photomicrographic camera (see Fig. 154), which is essential for examining fresh live liquid

Fig. 154. Vertical-bellows photomicrographic camera with illuminator.

specimens, is mounted rigidly on a heavy metal stand. The bellows can be raised or lowered along slides attached to the stand and locked in position at the desired magnification. The entire locked camera assembly can then be raised or lowered over the microscope without disturbing the bellows setting. The stand may have a large, heavy metal base on which the microscope, the illuminator, and the other accessories can be securely clamped in position after they are properly aligned.

HORIZONTAL-BELLOWS CAMERA

Although a vertical-bellows camera is usually preferred to one designed for use in the horizontal position, the horizontal camera does offer certain advantages. The microscope, the illuminator, the camera, and any accessories can be rigidly mounted on an optical bench or other precisely level surface, so that each piece of equipment is in the exact line of the optical axis. With such an assembly, it is a simple matter to move the camera or the illuminator backward or forward, or to exchange one piece of equipment such as the light source for another, and still have all components exactly aligned.

ILLUMINATING THE SPECIMEN

Illuminators

For several reasons, the light source used for photomicrography must provide more intense illumination than that used for visual microscopy. First, intense illumination allows reasonably short exposure times. Furthermore, color filters are extensively used in photomicrography, and these filters cut down the illumination. Finally, adequate illumination is necessary to focus the specimen precisely on the ground-glass screen.

The carbon arc lamp (which is described on page 31) is the type of illuminator most commonly used for photomicrography. The arc lamp is especially suited to high-power work.

Incandescent light sources, such as the tungsten ribbon-filament lamp, are also used as good sources of homogeneous, more highly concentrated light. Their chief advantage is that the position of the lamp need not be frequently adjusted in order to maintain uniform illumination of the field. These lamps are available with variable resistances to regulate light intensity.

Photoflood lamps are occasionally used, generally being coupled to a resistance in series to control intensity. Mercury- and sodium-vapor lamps are used when monochromatic light is essential.

A glass water cooling cell is usually placed between the field condenser of the illuminator and the microscope to cool the intense, concentrated light beams before they enter the microscope. If they are not cooled, their heat may soften the cement that holds the lens elements together and thereby permanently damage the optical parts. Also available are heat-absorbing glass filters, which are more convenient to handle and perform better. These filters are fashioned of a series of glass strips instead of one large piece, so that they will not be damaged by thermal stresses or cracking.

Methods of Illumination

The various methods of illuminating the specimen for visual microscopy that have been described can also be used in photomicrography. These include bright-field, dark-field, phase-contrast, polarized-light, oblique, ultraviolet, infrared, fluorescent, and monochromatic illumination. In each case, the procedure to be followed is similar to that previously described. Köhler illumination is the system best suited to photomicrography, since it gives the sharpest image and highest resolution possible. A detailed outline of the procedure for obtaining Köhler illumination appears on page 42.

OBLIQUE INCIDENT ILLUMINATION

Illumination of opaque objects for photomicrography with incident or reflected light is similar in many respects to the procedures described for transmitted light. Incident light with the vertical illuminator (see page 277), which gives flat light without shadows, is excellent for metallurgical or mineral specimens. In most cases, however, oblique incident illumination is used to achieve highlights and shadows in the specimens in order to produce a more attractive photomicrograph. The results obtained are similar to those in a conventional photograph.

Among the different procedures that may be used for obtaining oblique incident illumination are the following:

A. A spotlight from a single source, mounted on a stand that can be adjusted to the desired height, is positioned above the specimen at an oblique angle and focused on it (see Fig. 155). The illuminated

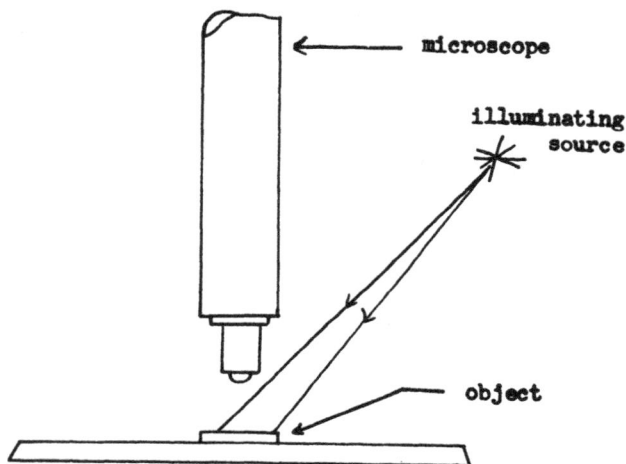

Fig. 155. Diagrammatic representation of oblique incident illumination with a spotlight (discussed in the text).

image can be seen on the ground-glass screen of the camera. The position of the light may be changed so that the best image is achieved without an undue amount of contrast or too deep shadows.

B. Several tungsten lamps of different intensities are positioned above and on opposite sides of the opaque specimen to give unbalanced diffused illumination. Beautiful photomicrographs of opaque live specimens can be obtained with this procedure.

C. A beam from an intense carbon arc lamp or zirconium arc illuminator is projected at right angles to the optical axis of the microscope and deflected to illuminate the specimen in one of several ways. This procedure is used mainly for high-power work, using an oil-immersion objective, although it can also be used advantageously with lower magnifications.

One way of deflecting the beam is to place a small concave mirror on the opposite side of the specimen and in the same plane as the beam. The beam passes above the object, strikes the mirror, which is positioned at an angle, and is reflected onto the specimen at a suitable oblique angle (see Fig. 156). A frosted glass may be used in conjunc-

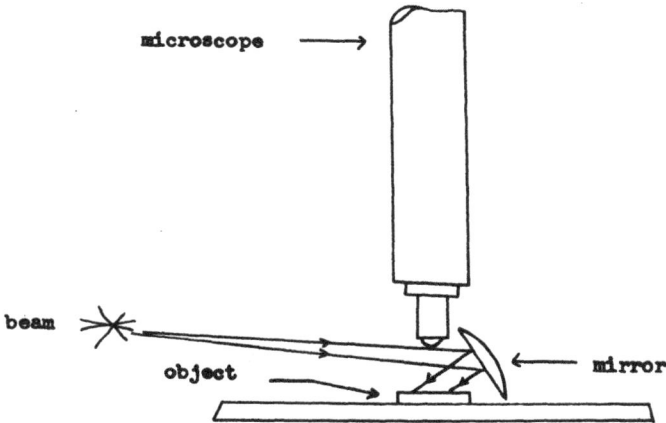

Fig. 156. Diagrammatic representation of oblique incident illumination with a spotlight and mirror (discussed in the text).

tion with the mirror to soften the light. The frosted glass is placed on the microscope stage, and the specimen is placed on this glass.

Another method of deflecting the beam is to use a small 90° prism instead of the small mirror. The prism is placed in the same plane as the beam and deflects the light before it reaches the specimen. The prism is so positioned that the light rays are deflected down onto the specimen (see Fig. 157).

It must also be noted that colored filters, polarizers, and other optical devices can be placed in the path of the light beam to achieve different results in photomicrography.

Colored Filters

Filters of various colors are used in photomicrography to achieve certain effects. These filters absorb the unwanted components or

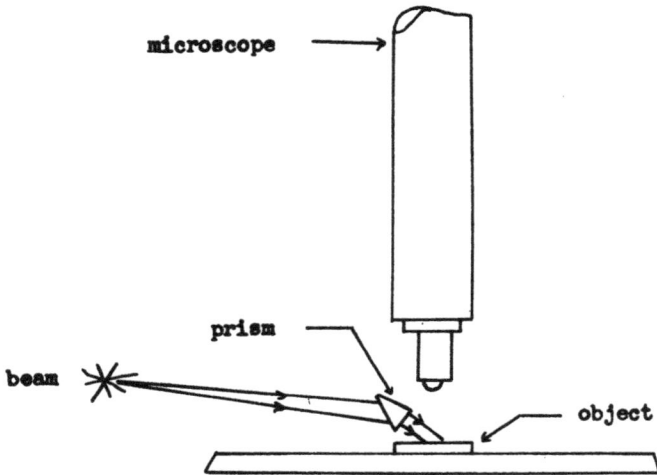

Fig. 157. Diagrammatic representation of oblique incident illumination with a spotlight and prism (discussed in the text).

wavelengths from white light as it passes through and transmit an approximate monochromatic light consisting of a narrow range of wavelengths of the color desired.

In black-and-white photomicrography, colored filters are used to make the colored details in a specimen appear either lighter or darker on the film, to increase the contrast of the specimen, and to achieve greater resolution of fine details. Generally, achromatic objectives give better resolution with monochromatic light. If the filter used is the same color as the specimen, the details of the specimen generally appear in lighter tones against a darker background. If the filter is complementary in color to the specimen, the details of the specimen appear in darker tones against a contrasting lighter background.

Colored filters are also used extensively in photomicrographic work to transmit monochromatic light of the wavelengths for which the objective lenses of the microscope have been chromatically corrected. Because achromatic objectives produce best results with green light, a green filter is recommended for routine work.

It is possible to isolate a narrow region of monochromatic light by

using colored filters singly or in pairs. It may be necessary to experi-
ment with many different filters to determine which filter or filters
give the best results. The following list can serve as a rough guide
for selecting the correct contrast filter:

Color	Filter
violet	orange
purple	yellowish-green
blue	yellow
green	red or pink
yellow	blue
orange	blue or violet
red or pink	green

Illuminating the specimen with light of a shorter wavelength generally gives greater resolution. When using a blue filter, a mercury-
vapor lamp, or a carbon arc lamp, it is essential to remove any
ultraviolet radiation that may be present, since achromatic or apochromatic objectives are not corrected for ultraviolet and the photomicrograph will not be sharp and crisp. A special ultraviolet-
absorbing filter, such as a Kodak Wratten Filter No. 2B, is used in
conjunction with these illuminators to stop the transmission of any
ultraviolet light.

Among the various types of filters used in photomicrographic
work are the following:

1. Permanent, durable all-glass filters.

2. Dyed gelatin film squares, which are mounted or sandwiched
between glass plates (Wratten filters). These filters are widely used
in microscopy.

3. Colored liquids, which are placed in the water cooling cell to
serve as both filter and coolant. For example, a liquid cooling cell
containing a solution of sodium nitrite (about 1%) is used to absorb
the heat from the illuminator and to remove any ultraviolet radiation.

4. Interference filters, which are made by depositing very thin
layers of metallic materials on glass in a vacuum. These filters pro-

duce interference effects and can yield very narrow bands or wavelengths of monochromatic light in the visible spectrum. Interference filters are superior to glass filters not only because they give more selective transmission of a monochromatic color, but also because they have rather small absorption losses and therefore give high light transmission.

5. Neutral-density filters, which are used primarily to reduce the intensity of illumination so that direct viewing can be done without injury to the eyes. They are also used to increase the exposure time of the film when necessary. These gray-glass or gelatin-film filters are produced in different optical densities that transmit only a certain amount of the incident light. For example, Kodak Wratten neutral-density filters are made in different degrees of opaqueness to transmit from 1% to about 80% of the incident light.

A filter of a given color is useful only with a film that is sensitive to light of the color transmitted by that filter. Yellow filters, for example, can be used only with orthochromatic film, since this film is sensitive only to green and blue light. A wide range of colored filters may be used to achieve specific effects with panchromatic film, which is sensitive to almost all visible colors as well as to ultraviolet.

For either black-and-white or color photomicrography, a good general rule is to select the type of film and light filter that will accurately record the best and sharpest specimen image. One method of selecting the correct filter is to examine the specimen under the microscope using one filter or a combination of filters, choosing the filter or filters that give the effects desired. Tables 6 and 7 will be found helpful in making the initial selection.

TABLE 6.   KODAK WRATTEN FILTERS FOR MICROSCOPY*

No.	Visual Color	Spectral Transmission
25	red	590 nanometers into the infrared
58	green	480–620 nanometers
47	blue	370–510 nanometers
35	purple	320–470 nanometers and 650 nanometers into the infrared
22	yellow-orange	550 nanometers into the infrared
29	deep red	610 nanometers into the infrared
15	deep yellow	510 nanometers into the infrared
45	blue	430–540 nanometers
11	yellowish-green	For correct tone reproduction with tungsten light

TABLE 7.   KODAK WRATTEN VISUAL "M" FILTERS FOR
PHOTOMICROGRAPHY*

No.	Color	Use
78AA	blue	Photometric filter; can be used to convert the color quality of light from incandescent tungsten lamps of the common type to that which is approximately visually equivalant with their commonly accepted standard daylight appearance.
38A	blue	Increases apparent contrast in faintly stained yellow or orange preparations. Helps in resolution of fine detail.
45A	blue-green	Especially useful for obtaining highest possible visual resolving power, required in fine work such as study of diatom structure. Has no red transmission; dominant wavelength range is 470–480 nanometers.
66	light green	Contrast filter for use with pink- and red-stained preparations. Preferred by some workers for general use in place of No. 78AA.

58	green	Contrast filter; used with faintly stained pink or red preparations.
15	deep yellow	Increase contrast in blue preparations.
22	yellow-orange	Facilitate observation of detail in insect mounts by reducing contrast between preparation and background.
25	red	Contrast filter; used with preparations stained with such stains as methylene blue and methyl green.
96	neutral	Reduces intensity of illumination. Neutral density (1.0) transmits 10% of incident light.

*Courtesy Eastman Kodak Co.

FOCUSING THE SPECIMEN

The image of the specimen must be brought into sharp, critical focus in order to obtain a sharp photomicrograph. With low magnifications, the center portion of the projected image can be visually focused on the ground-glass screen. It is more difficult to focus small details or fine structures, however, since the texture of the ground surface is coarse and will cause the structures to appear indistinct. When a bellows camera is used, therefore, a magnifying lens is used in most cases to aid in focusing. If an eyepiece-attachment camera is used, focusing of the specimen is done through the observation eyepiece.

To aid in critical focusing of the specimen when using a bellows camera, a ground-glass screen with a clear center portion is used. These are commercially available, but they can also be made in the following manner: With black India ink, draw two diagonal lines on the ground surface of the glass screen to locate the center. Put a drop of Canada balsam or other mounting medium at the intersection of the lines. Center a round cover glass over this drop, press gently, and allow to dry. The result will be a permanent clear window in the center of the ground glass.

When the image is to be brought into focus, place a magnifier or magnifying lens over this window and focus on the diagonal lines. Bring the image into approximate focus on the ground-glass surface. Then look through the magnifier positioned over the clear window and bring the image into sharp focus.

Another method is to move the magnifying lens slowly back and forth across the diagonal lines.   When the image is in critical focus, there will be practically no movement of the image in relation to the lines.

DETERMINING TOTAL MAGNIFICATION

It is essential to know the enlargement, or total magnification, of the specimen in order to evaluate it more fully in relation to its original size.   When photomicrographs are taken or the specimen is viewed on a ground-glass focusing screen, the total magnification depends on the distance between the eyepiece and the screen and the total magnification power of the eyepiece and objective.

The total magnification of the eyepiece and objective is determined at a standard distance of 250 mm (about 10 in.) from the eyepiece, since the normal viewing distance with the eye is 10 in.   If the distance of the ground-glass screen from the eyepiece is greater than this, the magnification of the image will be greater.   To determine the added magnification, this distance in millimeters is divided by the standard distance of 250 mm.   Thus, the total magnification of a projected microscope image can be determined by the formula $M = DM_1/250$, in which M is the total magnification of the projected image; D is the projection distance in millimeters; and M_1 is the total magnification of both objective and eyepiece, based on a microscope tube length of 160 mm.

For example, if the total magnification of the objective and eyepiece is 100X and the projection distance is 750 mm, the magnification of the projected image is calculated from the formula as follows: $M = (750 \times 100)/250 = 300X$.

If it is necessary to determine the magnification with extreme accuracy, a stage micrometer, scaled accurately in 0.01-mm divisions, is brought into focus on the ground-glass screen. The distance between the lines on the focusing screen can be easily measured with a pair of dividers. The exact magnification can then be determined by simple arithmetic. The magnification thus calculated will be valid only for the specific combination of objective, eyepiece, and bellows extension used. This procedure can also be followed for eyepiece-attachment cameras if focusing is done on a ground-glass screen.

PHOTOGRAPHIC PRINCIPLES AND PRACTICE

In order to make good photomicrographs, the microscopist must have a basic understanding of the theory and practice of photography. The supplies and equipment necessary for photomicrography, for example, are essentially the same as for conventional photography. Furthermore, although the processing of color photomicrographs is usually done commercially because it is rather complicated, the individual worker can process most black-and-white and toned photomicrographs in his own darkroom better than commercial processing firms, since he can make more extensive adjustments in processing conditions.

Photographic supplies, equipment, and procedures, as they apply specifically to photomicrography, are discussed in some detail in the following paragraphs.

Selecting Film for Black-and-White Photomicrography

Either sheet film or glass plates may be used for black-and-white photography. The film (or plate) consists of a thin, transparent sheet of cellulose acetate or other plastic (or glass) coated on one side with a thin layer of a gelatin emulsion in which crystals of silver bromide, silver chloride, or silver iodide are suspended. Exposure of these silver salts to light or to invisible radiation such as ultraviolet or infrared reduces them to metallic silver. The amount of reduction of the silver salts depends on the intensity of visible or invisible radiation striking the film emulsion. Brighter details of the specimen will cause a greater reduction of the silver salts; darker areas, a lesser reduction.

After exposure, the exposed film is developed in various chemical solutions. The developed film or plate is called a *negative* because of the inverse relationship between the brightness of a given part of the specimen and its brightness in the negative — the brighter a given area, the darker the corresponding area of the negative, and vice-versa.

Sheet film and glass plates are available in wide variety. The choice between film and plates is one of personal preference.   Glass plates, in addition to having all the necessary characteristics for producing a good negative, are rigid, and thus give greater flatness and dimensional stability than sheet film, which is flexible. Sheet film is nevertheless preferred by the majority of microscopists, since it requires less storage space and is not subject to breakage.

FILM CHARACTERISTICS

In selecting sheet film or glass plates for black-and-white photography, the following important film characteristics must be considered: (1) color sensitivity, (2) emulsion speed, (3) grain size or resolving power, (4) contrast range, and (5) exposure latitude. These characteristics are discussed separately below.

COLOR SENSITIVITY

The color sensitivity of the film emulsion is one of its more important characteristics, since it determines the quality of the photomicrograph, whether colored filters are to be used, and how the film is to be safely processed in the darkroom. Three types of emulsions with different sensitivities to color are commercially available.

Orthochromatic film emulsion is sensitive to the blue, green, and ultraviolet regions of the spectrum and is generally used to photograph specimens in which red is not the predominant color.

Special orthochromatic film is manufactured for metallographic use. This film is fine-grained and has degrees of contrast that are highly desirable for this work.

Panchromatic film emulsion is sensitive to all visible colors of the spectrum and to ultraviolet light. This film is further subdivided into two categories according to its sensitivity to certain colors: *Type B* is especially sensitive to green, its sensitivity to this color being almost equal to that of the eye. *Type C* is especially sensitive to red.

Panchromatic film is used to photograph colored specimens. Filters can be used with this film to control the intensity and the reproduction of each color in the film.

Blue-sensitive film emulsion, also called *ordinary* emulsion, is sensitive only to the shorter wavelengths of light, such as blue, violet, and ultraviolet. Ordinary film is used for special purposes, such as when the specimen is illuminated with ultraviolet light.

EMULSION SPEED

The emulsion speed of a film determines the amount of light and the exposure time required to photograph the specimen. The higher the film speed, the less the light or the shorter the exposure time; the lower the film speed, the more the light or the longer the exposure time. High-speed film is rarely needed for photomicrography. In fact, since the specimen is motionless in most cases, long exposures are preferred because exposure time becomes less critical as it increases. Furthermore, high-speed films gain speed only by sacrificing other desirable film characteristics, such as fine emulsion grain size and latitude of contrast. A high-speed emulsion may be of some value for photographing live specimens, but film of medium speed gives satisfactory results for most purposes.

GRAIN SIZE OR RESOLVING POWER

Films or plates in which the silver salts are more finely dispersed in the film emulsion, a property known technically as *fine grain size*, produce sharper negatives with good image quality and greater resolving power. Although the resolution of the specimen image depends primarily on the optical arrangement of the microscope and the specimen, fine-grain film aids in achieving optimum resolution. Fine-grain film is also advantageous when the negative image is to be further enlarged, since the quality of the enlarged image is not degraded by visible grains in the photograph.

CONTRAST RANGE

To achieve a good photomicrograph, it is essential that there be substantial contrast between light and dark areas of the specimen. A negative with little contrast (that is, little gradation of the reduced black silver within either the light or the dark areas) appears uniformly black or light, and the print made from it appears flat.

Although a type of film that can produce negatives with different degrees of contrast is available, the contrast of the negative depends primarily on the developer used and the developing time. For example, a film with high-contrast characteristics will show low or medium contrast if it is developed either for a shorter time than prescribed or in a dilute developing solution. The type of developer, concentration of the developing solution, and development time are therefore important factors in achieving maximum contrast.

The film to be used, whether sheet or plate, should be selected first for the contrast desired, which depends on the brightness range of the microscopic image, and then for its sensitivity to the kind of illumination to be used. The processing information that usually accompanies the film should be strictly followed to achieve the best contrast.

EXPOSURE LATITUDE

Films or plates with wide latitude in exposure time give satisfactory negatives even if exposure time is not precise, since they allow greater tolerance of exposure.

Selecting Film for Color Photomicrography

In photomicrographic work, it is often necessary to record the artificially produced color contrast of a stained specimen or the natural colors of a colored, unstained specimen. Color films are used for these purposes, although the colors recorded are in many cases either too intense or not true and are thus not comparable to those of the specimen. Care must therefore be taken in making color photomicrographs or in assessing those in which the color rendition is not faithful.

There are two types of color sheet film—*reversal* and *negative*. Reversal film produces positive transparencies, which are usually viewed by means of a projector. Color prints, however, can be produced from separation negatives made from these positive transparencies, or by using the three-color filter and dye process. Negative film produces negatives in which the colors are complementary to the actual colors of the specimen. Color prints or transparencies can be made from color negatives by contact or projection printing, with the use of compensating filters and special printing materials in processing.

Determining Film Exposure Time

In order to obtain excellent photomicrographs, it is essential to determine the correct exposure time for the film or plate in use. In general, the more intense the illumination of the projected image and the greater the emulsion speed, the less the exposure time required. The intensity of illumination of the projected image depends on the total magnification power of the objective and eyepiece, the intensity of the light source, the aperture opening of the iris diaphragm, and, finally, the distance of the film or plate from the microscope eyepiece.

The usual way of determining exposure time is by trial and error. For black-and-white photomicrography, a series of test exposures are made of the projected specimen image, and the one giving the best reproduction of detail in the darkest portion of the specimen is used. The same procedure is followed for color film, and the correct exposure time is the one that reproduces small details in the brightest portion.

The procedure for determining exposure time by trial and error is as follows:

Step 1. Illuminate the specimen placed on the microscope stage and bring it into sharp focus.

Step 2. Position the camera over the eyepiece and focus the image on the ground-glass screen. Set the shutter for time exposure.

Step 3. Load the plate-holder with film and then replace the ground-glass screen with the plate-holder. Close the shutter.

Step 4. Uncover the film completely by removing the dark shield from the plate-holder, and expose the film for 3 seconds. Slide the dark plate-holder shield back to cover a portion of the exposed film or plate, approximately 1 in., and repeat the 3-second exposure of the portion still uncovered. Slide the dark shield over another 1-in. portion of the film surface and expose again for 3 seconds. Continue this procedure until the last remaining portion of the film plate is exposed and the shield completely covers the film.

Step 5. Develop the film and examine the negative. Choose the portion that gives the best reproduction of the specimen, that is, the part that gives an image with good density.

The correct exposure time for this film or plate will thus be 3 seconds or a multiple thereof, depending on the portion chosen. This time will be valid only for taking photomicrographs under exactly the same conditions: light intensity, specific combination of objective and eyepiece, eyepiece-bellows distance, magnification, type of film, and processing procedure. Since there are many variables involved in the photographic process, it is advisable to prepare a new test exposure film strip for each new specimen.

The exposure time can also be determined with exposure meters, which register different intensities of light and are calibrated for films of different speeds. The following factors must be considered in determining exposure time with a meter: (1) the exact distance between the exposure meter and the film; (2) the type of light source to be used for illuminating the specimen; and (3) the emulsion speed, or film speed; and (4) the method of developing the film. In addition, the meter reading must be taken without the use of light filters.

A colored filter usually cuts down the amount of illumination reaching the film emulsion. The manufacturer usually specifies the *filter factor*, a number that indicates the amount of additional light that must be used to give the correct exposure. The filter factor can be easily ascertained by test-exposing to obtain two negatives of equal optical density, one with and one without the filter, with all other conditions equal.

Selecting a Photographic Printing Paper

Photographic printing papers are manufactured especially to produce a positive image from exposed negative film. These papers are similar to film in that they are coated with a light-sensitive silver emulsion. When printing paper is placed in contact with a negative and exposed to light, or when light is projected through the negative onto the paper, a latent positive image is formed. This image is then made visible by chemical means.

Printing papers for black-and-white photography are available in different grades to accommodate negatives of varying densities. They are also produced in single- and double-weight stock and in a variety of surface textures (glossy, matte, and semimatte).

Glossy papers are most often used for photomicrographic work, since they reveal maximum detail. They are available in various sizes and in different grades of contrast, the latter designated by numbers that denote papers of contrast gradually increasing from low to high—*0* (extra soft), *1* (soft), *2* (normal), *3* (hard), *4* (extra hard), and *5* (very hard). The contrast of a photomicrograph can be increased or decreased, within narrow limits, by selecting the appropriate grade of paper. This is done by examining the negative in the bright light and noting the degree of contrast, not the actual negative density.

Selection of a paper with the proper contrast may also improve somewhat the results obtained from negatives that are faulty because of under- or overexposure or under- or overdevelopment. To achieve the best photomicrographs, however, correct exposure time and correct development of the film are essential.

For properly exposed and developed negatives, a medium-contrast paper (No. 2 or 3) is used. For underexposed negatives that lack detail but have been developed normally, a high-contrast paper (No. 4 or 5) is used. A high-contrast paper is also used for overexposed negatives that have been developed normally, since such negatives appear dense and lack contrast. Underdeveloped negatives — that is, films that have been properly exposed but have been developed for less than the allotted time — also lack contrast and must therefore be printed on a high-contrast paper. Overdeveloped negatives that have been correctly exposed would appear dense with high contrast. For such negatives, a low-contrast paper (No. 0, 1, or 2) is used.

Darkroom Equipment

Since film and printing papers are sensitive to light, they must be developed and printed in a darkroom. In the darkroom, use is made of various types of colored lights, or *safelights* (see Fig. 158), which

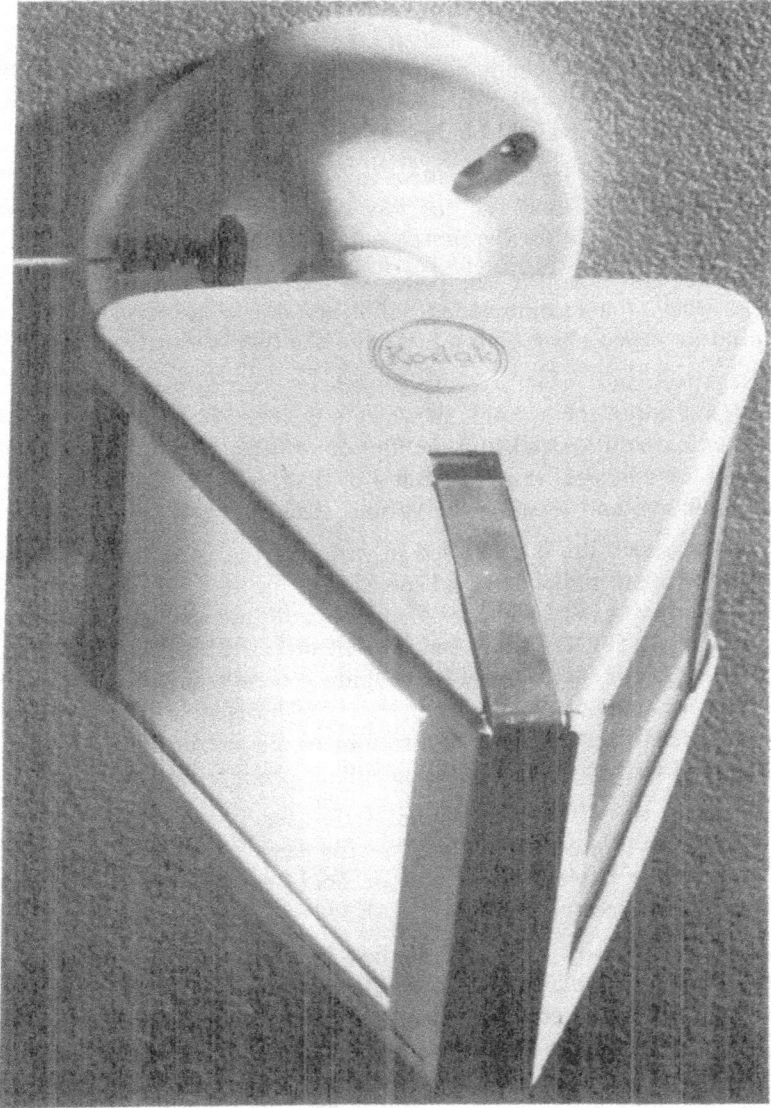

Fig. 158.   Safelight.

do not affect the plates, film, or printing papers but do provide adequate illumination for the darkroom worker.

A deep red or ruby subdued light may be used for orthochromatic film. Since panchromatic film is rather sensitive to red light, a special green safelight may be used for this film—lighter green for the slower type, darker green for the faster. For best results, however, panchromatic film should be developed in complete darkness.

Printing papers are less sensitive to light than film, so that a yellow safelight can be used for the less sensitive contact papers and an orange-red safelight for the more sensitive bromide enlarging papers.

Neither film nor printing paper should ever be exposed to white light. Only the recommended safelights should be used, and they should be kept at least 3 to 6 feet from the light-sensitive material.

Negatives and prints are developed in plastic, metal, or enamel *developing trays*, the bottoms of which are ribbed to facilitate the flow of chemical solutions around the film or printing paper. These trays are available in sizes ranging from 4 by 5 in. to 20 by 24 in. for developing, fixing, and washing the various sizes of negatives and prints.

Roll or sheet film is processed in various types of *developing tanks*. Plastic or metal tanks designed specifically for roll film are provided with spiral reels onto which one or more rolls are loaded for processing (see Fig. 159). Plastic, metal, or vulcanized rubber tanks designed for sheet film or plates (see Fig. 160) are generally supplied with film hangers for sheet film and processing racks for plates. These hangers or racks eliminate handling of the film or plates during developing, fixing, and washing. The tanks have light-proof covers.

Plastic or metal film-washing tanks are used for thorough washing of the film. Water is circulated in the tank by a perforated metal tube or other device, and a slot or hole near the top of the tank functions as a safety overflow.

Large batches of prints are washed in a special print washer, a large tank into which fresh water flows from below under sufficient pressure to ensure constant agitation of the water surrounding the prints. A siphon mechanism attached to the side of the tank maintains a constant water level and discharges the excess, contaminated water into a waste receptacle or sink. The print washer works automatically, without special attention.

Fig. 159.   Roll-film developing tank with reels.

A *timer* (see Fig. 161) is necessary when processing film and making prints.   There are many types of timers, both mechanical and electrical.   A suitable timer is one in which the setting, in minutes or seconds, actuates both the alarm and the clock mechanism.   Additionally, it should be possible to stop and restart the timer without resetting the hands to zero.

Since correct temperature of the various chemical solutions is essential to obtaining optimum negatives and prints, a *thermometer* must be used to determine the temperature of these solutions during the processing of the film and paper.   Various types of thermometers designed for use in photographic processing are available.

An *enlarger*, used in printing by projection (enlargement), is actually a camera in reverse.   The negative is illuminated and the image then projected through the enlarger lens onto the printing paper.

Fig. 160.   Sheet-film developing and fixing tank.

There are many types of enlargers available.   The vertical type is preferable for photomicrography, and it should accommodate negatives from 35 mm to 4 by 5 in.

Fig. 161.   Timer.

Printing frames are used in the contact method of making prints. They are available in wood or plastic and in various sizes.   *Easels* and *masking frame*s are used in the projection (enlarging) method of making prints. The easel accommodates various sizes of printing papers by means of adjustable masks attached to it. In this way, the paper is held rigidly during exposure and unwanted portions of the negative are masked out.

Highly polished stainless-steel or chrome-plated steel *ferrotype plates* are essential for making glossy prints. Special ferrotype machines that can process a large quantity of prints are also available. The ferrotype plate, which may be either flat or cylindrical, is heated electrically to hasten drying and ferrotyping of the prints.

A *print roller* or *squeegee* is used to remove the excess moisture from prints while ferrotyping. Rollers and squeegees are available in various sizes, and a size slightly larger than the width of the ferrotype plate should be used.

The *print trimmer* is a labor-saving device for trimming prints with even borders and for cutting printing papers and mounting boards to exact size. The guillotine type trimmer is the one most commonly used. It consists of a flat wooden mounting board, on which the paper is laid, and a sharp guillotine knife attached to the board that is brought down to cut the paper.

Record cards, for recording the data for each negative, should be readily available. The information on the cards should include: date, negative number, subject, total magnification, type of illumination, magnification of objective and eyepiece, filter used, exposure time, remarks, and any other data deemed necessary.

Processing Sheet Film and Glass Plate Negatives

After the sheet film or glass plate has been exposed, it must be processed to reveal the latent image formed by reduction of the silver salts during exposure. This processing, which produces the negative, consists basically of four steps—*developing, fixing, washing,* and *drying*. The general procedure for each of these steps is described below.

DEVELOPING

Developing of film in the appropriate chemical solutions completes the selective reduction of silver salts to silver in those portions of the negative that were affected by light during exposure. Essentially, developing reveals the latent image in the negative.

The standard developing solutions for black-and-white film contain four major chemical ingredients: a *reducing chemical*, which further reduces the exposed grains of silver salt to metallic silver; an alkaline *activating chemical*, which activates and expedites the reducing process; a *retarding agent*, which inhibits excessive action of the activator; and a *preservative*, which prevents deterioration of the developer.

The chief requirements for the reducing chemical in a developer are that it have selective action and that this action be controlled. Some chemicals that have been found to be effective reducing agents are pyrogallol, metol or elon, hydroquinone, and amidol.

When any of these chemicals is dissolved in water, an alkali, such as sodium carbonate, must be added to speed up the developing process and aid in increasing the contrast in the film image. It is essential that the correct proportion of alkali be added, since an excess will cause the reducing chemical to reduce not only the portions of the negative that were exposed to light, but also those portions that were not; such indiscriminate reducing action will produce fogged negatives.

In order to inhibit the action of the alkali, another chemical, potassium bromide, is added to the developer.

Finally, sodium sulfite is added to the developer as a preservative. The reducing chemical, being a strong reducing agent, has a great affinity for the oxygen in the air. The sodium sulfite combines with the reducing chemical to form a complex salt, which is far less reactive with oxygen, and thus prevents the reducing agent from taking up oxygen and thereby becoming inactivated. Sodium bisulfite and sodium meta-bisulfite are also used in place of sodium sulfite.

The type of developer selected depends on the type of film and the contrast and density desired in the processed negative. Developing time also affects the contrast, but there is a limit to the degree of increased contrast that can be obtained with a specific developer by prolonging developing time. For best results, the developer formula and developing time recommended by the film manufacturer should be followed.

The time necessary for developing glass plates or sheet film can be determined by either of two methods.

In the first, the film is observed under a safelight in the darkroom as it is being developed, and development is stopped when the negative is considered to have the correct density and contrast. This method requires considerable experience.

The second method relies on the use of a standard formula for the developing solution and on adherence to the manufacturer's instructions for the correct solution temperature and developing time necessary to produce a negative with the right contrast. This method does not allow compensation for any errors in exposure or for any deficient contrast of details in the specimen. The optimum solution temperatures and developing times for specific degrees of contrast have been determined empirically, and these data are usually packaged with the film by the manufacturer.

Slight correction of overexposed film or plates can be made by underdevelopment — that is, by developing for a shorter time than is recommended by the manufacturer. Underdevelopment will not correct any initial lack of contrast in the negative caused by incorrect exposure, even though the density of the negative is reduced; underdevelopment, in fact, further reduces the original contrast. Similarly, underexposed film or plates can be corrected, but only slightly, by overdevelopment. Negative contrast will increase slightly, but the reduced silver areas will be less dense and detail will be lost. Given these limitations and the possibility of further error, the best rule, even with overexposed or underexposed film, is to adhere strictly to the manufacturer's recommendations.

Temperature significantly affects developing time, which decreases as the temperature increases. The temperature of the developing solution should fall within the range of 65–70°F and should be nearly constant throughout. Lower temperatures slow development and increase the time necessary. Furthermore, each constituent of the developer reacts differently to any radical change in temperature, disrupting the balanced action of the formula. At higher temperatures, the reducing chemical oxidizes more readily, discoloring and weakening the developer. Such a solution will stain negatives and lose its effectiveness more quickly. The film emulsion also softens at higher temperatures.

FIXING

The next step after developing is fixing, the purpose of which is to remove any unexposed silver salts from the film emulsion, leaving the negative image formed by the reduced metallic silver. The fixing bath contains a chemical that reacts with the silver salts but does not affect in any way the gelatin film emulsion or the silver image. Sodium thiosulfate (commonly called *hypo*) is the chemical most commonly used for this purpose.

Prolonged immersion of film in the hypo may discolor the fixer. This discoloration is caused by oxidation of any developer carried over in transferring the film from developer to fixing bath. The discoloring reaction may not only stain the negative, but may also, if the fixing action is prolonged, reduce the density of the negative image somewhat and soften the gelatin film emulsion. Other ingredients are therefore added to minimize these undesirable effects and increase the efficiency of the fixer. Acetic acid is added to neutralize the alkali in any developer carried over into the fixer. Sodium sulfite may be added to prevent any premature oxidation of the fixer and thus prevent staining of the film or prints, as well as to prevent any reaction between the acetic acid and the hypo. Potassium alum is added to harden the gelatin emulsion of the film and print.

The higher the concentration of the hypo, the less will be the time required for fixing. A fixing solution that is weak or nearly exhausted will require a longer fixing time. The thickness of film emulsion layer also affects the time required for fixing, so that more time must be allowed for fixing thicker film or plates. Agitating the solution serves to accelerate the fixing action.

WASHING

When the film has been fixed for the proper length of time, all the remaining chemicals with which the negative is saturated must be removed. To do this, the negative is transferred from the fixing bath to a tray, tank, or spiral washer and washed thoroughly in running water for about 30 minutes. The fixer must be completely removed, since any remaining traces will eventually discolor the negative badly.

As a final rinse, the negative should be washed with water containing a few drops of a wetting agent, which will eliminate any tendency of the excess water to form water marks during the drying process.

DRYING

For drying, the wet negative is placed in a rack or suspended from clips in a dirt-free atmosphere at room temperature until thoroughly dry. A fan may be used to hasten drying.

TRAY METHOD OF PROCESSING

When trays are used for developing, fixing, and washing the negative, it is recommended that only one sheet or plate be processed at a time. The advantage of this method is that development can be observed under a safelight and the action stopped when the desired contrast is obtained. The tank method is preferred for processing more than one sheet or plate at a time.

The general procedure for processing sheet film or plates by the tray method is as follows:

Step 1. Arrange three trays of appropriate size, one each for the developing solution, acid-hardening stop bath, and fixing bath.

Step 2. Prepare the developing, hardening, and fixing solutions in the correct concentrations and pour each solution into a tray. Check the temperature of the solution.

Step 3. Set the timer for the correct developing time. Switch off the main light in the darkroom and switch on the correct safelight. If the processing is to be done in complete darkness, note carefully the position and content of each tray.

Step 4. Remove the sheet film or plate from the holder, place it emulsion side up in the tray containing the developer, and start the timer.

Step 5. Agitate the solution or move the tray slightly up and down throughout the developing period so that the film is evenly developed.

Step 6. Transfer the negative to the tray containing the acid-hardening solution and allow it to remain for about 2 minutes.

Step 7. Transfer the negative to the tray containing the fixing solution and allow it to remain for about 12 minutes with intermittent agitation. The main light may be switched on after the negative has been in the fixer for about 5 minutes.

Step 8. Transfer the negative to a washing tank and wash it thoroughly in running water for about 30 minutes. As a last rinse, transfer the negative to a water bath containing a few drops of a wetting agent and allow it to remain for about 1 minute.

Step 9. Remove the negative from the wash water and hang it up to dry in a dust-free atmosphere at room temperature.

TANK METHOD OF PROCESSING

The general procedure for processing sheet film or plates by the tank method is as follows:

Step 1. Dissolve sufficient developer in the necessary amount of distilled water and transfer to the developing tank. Check the temperature of the solution. The ideal temperature range is 65–75°F.

Step 2. Set the timer for the correct developing time. Switch off the main light in the darkroom and switch on the correct safelight for the film used.

Step 3. Remove the sheet film (or plates) from the holders. Attach the sheet film to the film hangers (or place the plates in the plate rack). Take care to separate the sheets or plates sufficiently, so that uniform development is not hindered and the emulsion surfaces are not scratched.

Step 4. Place the film hangers or plate rack in the tank containing the developer and start the timer. Agitate the solution around the film for at least 1 minute to wet the film or plates evenly and remove any air bubbles adhering to the film surfaces. This can be expedited by lifting the negatives from the tank during the first minute of agitation, allowing the solution to drain back into the tank, and then returning the film to the tank, repeating the procedure several times during development.

Agitate the solution for 30 seconds every 2 minutes during development.

Step 5. When development is complete, quickly remove the negatives from the developing tank and place them in an acid-hardening bath for about 2 minutes.

Step 6. Transfer the negatives to the fixing bath. Agitate the solution for about 1 minute and then allow the negatives to fix for about 12 minutes.

Step 7. Transfer the negatives to the washing tank. Turn on the main light in the darkroom. Wash the negatives in running water for about 30 minutes. As a last rinse, transfer the negatives to water containing a few drops of a wetting agent and allow them to remain for about 1 minute.

Step 8. Remove the negatives and allow them to dry in a dust-free atmosphere at room temperature.

Processing 35-mm Film Negatives

As previously mentioned, special plastic or metal tanks are available for processing 35-mm film. These tanks are equipped with a reel or spiral onto which the film is loaded and which then slips into the developing tank. A special light-tight cover with a central hole through which the various solutions are poured allows processing of the film in ordinary light. The processing procedure is as follows:

Step 1. Load the 35-mm film roll into the reel or spiral in total darkness, following the manufacturer's directions for loading. The usual spiral has an opening on either side through which one cut end of the film roll is pushed until the entire roll is loaded. Some spirals are provided with a ratchet device that is rotated or pushed back and forth to turn it and load the film. It is well to practice by loading some dummy 35-mm film in the light before attempting to load exposed film in the darkroom.

Step 2. Place the loaded spiral in the empty tank. Put the cover on the tank and rotate it clockwise to lock it in place. The main light may now be turned on. Some tanks have a special stirring rod that can be inserted through the central hole in the cover to stir the solutions during processing. Others are provided with a cap that fits securely over the central hole, so that the entire tank can be turned rapidly upside down and back again to agitate the solution.

Step 3. About 250 ml of solution is generally needed to fill a 35-mm processing tank. Pour the developer through the central hole, holding the tank at an angle to prevent any air pockets. Stir or agitate the solution for about 1 minute, so that the film is uniformly bathed by the developer. Repeat the stirring or agitation every 1/2 minute until development is complete. Empty the tank by inverting it over a sink.

Step 4. Fill the tank with water several times and discard.

Step 5. Fill the tank with the fixing solution and agitate periodically during fixing. Fix the film for about 15 minutes. Remove the cover and discard the fixing solution.

Step 6. With the film still in the tank, wash it in running water for about 30 minutes, agitating it periodically during the washing process. As a last rinse, add a few drops of a wetting agent.

Step 7. Remove the film from the reel and clip it to a hanger. Allow it to dry in a dust-free atmosphere at room temperature. Clean and dry the tank, spiral, and cover.

Negative Defects

Processed negatives may have certain defects, and it is essential to establish the causes of these defects so that they may be corrected in future processing. Some of the more common defects and their causes are discussed here.

An overly dense negative may be caused by overexposure or over-development. Another cause may be that the developing solution was too strong or its temperature too high. A dense negative usually lacks good contrast.

A thin negative may be caused by underexposure or underdevelopment. It may also be caused by the developing solution's being too weak or too cold. A thin negative lacks contrast, but not necessarily detail.

Uneven development produces areas of different densities in the negative. It may be caused by any of the following errors in procedure: (1) placing the film emulsion side down, so that the emulsion touches the tray; (2) failing to agitate the solution during development; (3) allowing plates or sections of film to touch during development; or (4) failing to mix the developer properly.

Stains on negatives may be caused by contamination of the developing solution, insufficient fixing, or use of exhausted fixing solution.

Miscellaneous defects in negatives include: (1) fogging, which may be caused by unsafe light falling on the film during development; (2) white crystalline deposits, which may be caused by the use of hard water for washing, by insufficient washing or by overacidity of the fixing bath; (3) water stains, which are caused by splashing the dried negative with water; (4) reticulation (a leathery appearance in the negative), which may be caused by the developer's being either too strong, too alkaline, or too warm; and (5) fingerprints, which result from handling the film with the fingers, rather than with wooden or plastic tongs.

Printing Negatives

After the negative has been processed, a print of it must be made on photographic printing paper. Printing paper is similar to film in that one surface is coated with a light-sensitive silver halide emulsion. Consequently, although the paper is not so sensitive to light as film, the methods of exposing and processing the paper are similar to those prescribed for processing film and plates. The function of printing paper is likewise similar to that of film, in that a *positive* latent image of the negative is formed on the light-sensitive surface of the paper when it is exposed to light. Development of the paper then reveals this positive image.

The general procedure in contact printing consists of placing the emulsion side of the printing paper in close contact with the emulsion side of the negative. This is done by inserting the paper and negative in a printing frame with the negative face up to the light. In the projection (or enlarging) process, the paper and negative are kept separated — the paper being held rigidly in an easel, the film being placed in a special carrier and positioned near the source of light. In both procedures, artificial light passes through the translucent negative and strikes the paper for a specific interval of time.

DETERMINING PRINT EXPOSURE TIME

Because printing papers are less sensitive to light than film, they have greater exposure latitude, but correct exposure is nevertheless important if fine prints are to be obtained.

The exposure time for a specific grade of paper depends on the type of paper, the density of the negative, and the intensity of the illumination. The exposure time can be manipulated by varying one or more of the variables described for the development of film. If the print is overexposed, it must be removed from the developer quickly to prevent it from becoming too dark, with consequent loss of detail and contrast. If the print is underexposed, however, it will remain too light no matter how long it is kept in the developer. Thus, there is little that can be done in printing to correct for underexposure of the paper.

STRIP METHOD

One method of determining the correct exposure time for printing, by either contact or projection, is similar to that described for determining the exposure time of sheet film or plates. The general procedure is as follows:

Step 1. Repeat the procedure for determining the correct film exposure as described on page 316, using a piece of cardboard instead of the shield to cover each segment of the paper.

Step 2. Process the paper, wash in water, and examine in bright light.

Step 3. Use the exposure that produces the print segment with the greatest detail and the proper gradation of black and white highlights. The background should be white or slightly gray.

EXPOSURE SCALE METHOD

Another method of determining exposure time, also for either contact or projection printing, makes use of the exposure scale made by Kodak, which consists of a film negative divided into 10 segments, each of different density and numbered to give the correct exposure (see Fig. 162). The procedure for determining exposure time for printing by each method is described below.

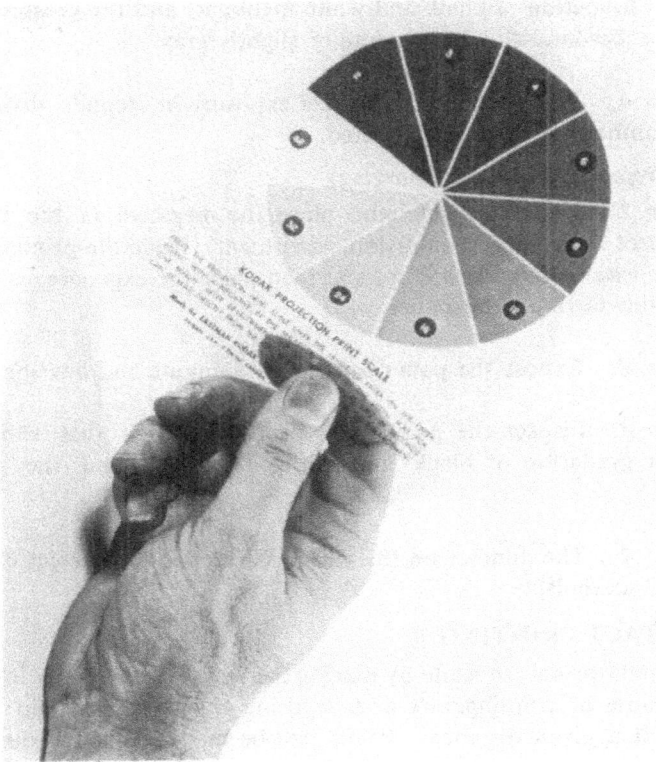

Fig. 162.   Print exposure scale.

For Contact Printing

Step 1. Place the negative emulsion side down on the emulsion side of the paper. Then place the exposure scale on the negative, emulsion side down.

Step 2. Fasten negative, paper, and scale securely in a printing frame with the exposure scale uppermost.

Step 3. Expose the paper for exactly 30 seconds.

Step 4. Develop, fix, and wash the print.

Step 5. Inspect the print and select the wedge that shows the proper gradation of black and white highlights and the greatest detail. The background should be white or slightly gray.

Step 6. To determine the correct exposure in seconds, divide by 2 the number on the wedge selected.

For Projection Printing (Enlarging)

Step 1. Under the safelight, place the negative in the negative carrier of the enlarger, emulsion side down. Place the printing paper on the easel emulsion side up, and then place the exposure scale emulsion side down on the paper.

Step 2. Expose the paper for exactly 1 minute and develop it.

Step 3. Inspect the print and select the wedge that shows the proper gradation of black and white highlights and the greatest detail.

Step 4. The number on the wedge selected is the correct exposure time in seconds.

CONTACT PRINTING

Contact prints are made by placing the negative and paper in a printing frame or printing box and exposing the paper to light from a lamp at a given distance. Prints can be made easily and quickly by this method.

If a white border is desired, a printing mask is placed between the negative and the paper. The mask is a piece of stiff black paper of the same size as the print, with a cutout portion the shape and size of the desired print area. This cutout may be a rectangle, a square, or a circle.

The printing paper is chosen according to the type of photomicrographs required. Glossy papers, however, are generally preferred.

The general procedure for contact printing is as follows:

Step 1. Open the back of the printing frame and remove it with its attached spring. Clean both sides of the glass window and place the frame on a flat surface with the glass side down.

Step 2. Place the negative in the frame emulsion side up. Take care in handling the negative not to mar it with fingerprints or other marks.

Step 3. Turn on the safelight and switch off any bright lights. Place the printing paper on the negative emulsion side down, so that paper emulsion and negative emulsion are in contact. Replace the back of the printing frame and affix the spring clamp, so that paper and negative are securely held in close contact.

Step 4. Invert the printing frame, so that the glass side is uppermost and facing the light.

Step 5. Set the timer and expose the negative and paper to a uniform light source for the proper time.

Step 6. Extinguish the light, remove the spring back from the printing frame, and remove the paper.

Step 7. Develop the paper under the same conditions as used for the test strip selected.

PROJECTION PRINTING (ENLARGING)

It must be noted that there is some loss of sharpness and definition in enlarged photomicrographs because of enlargement of the emulsion grains in the negative.

The general procedure for projection printing (enlarging) is described below.

Step 1. Insert the negative emulsion side down in the negative carrier of the enlarger and replace the carrier.

Step 2. Place a blank piece of white paper, of the same thickness and size as the printing paper to be used, in the easel. Adjust the easel and its masking frame to the required print size.

Step 3. Turn on the safelight and switch off the main light. Turn on the enlarger and adjust its height to obtain the desired enlargement. Open the lens diaphragm wide and bring the negative into sharp focus on the paper by means of the focusing knob. Use a magnifying glass on the paper to aid in focusing.

Step 4. Close the lens diaphragm down to the aperture used for the test exposure selected. (Closing the lens diaphragm down increases the depth of focus and may increase the image definition.) Switch off the enlarger. Remove the blank white paper from the easel and replace it with the printing paper, emulsion side up. *The easel must not be moved.*

Step 5. Set the timer and expose the paper.

Step 6. Remove the paper from the easel and develop the print under the same conditions as used for the test exposure.

DODGING

If illumination of the specimen during exposure was uneven, the negative may appear too dense in some portions. It is possible to give more light to these dense parts during the projection process by a procedure called *dodging*, which exposes the print more evenly and thereby produces a better photomicrograph.

Dodging can be conveniently done with an opaque light shield, such as a red plastic disc attached to a long, thin wooden or plastic handle. It is done by moving the shield continuously and rhythmically back and forth between the light source and the thin portions of the negative during exposure of the print. In this way, more light strikes the denser portions. The time and the speed of the back-and-forth motion of the light shield necessary for dodging can be determined only by trial and error.

Processing Prints

The procedure for processing exposed prints, which is decribed below, is similar to that for processing exposed negatives. It should be noted that there is wider latitude in developing printing paper than there is in developing film.

Step 1. Arrange three trays of appropriate size for the developing solution, fixing bath, and water bath.

Step 2. Prepare the developing and fixing solutions in correct concentration and pour each into a tray. Check the temperature of the solutions.

Step 3. Set the timer for the correct developing time. The safelight should be on when the print is transferred from the printer to the developing tray.

Step 4. Place the exposed print emulsion side up in the tray of developer. Rock the tray back and forth and from side to side to agitate the solution, so that fresh developer is continually brought to the emulsion surface and any air bubbles are removed.

The length of time necessary for development of the print can be determined either by test exposure, as described on page 332, or by watching the development under the proper safelight. If the latter method is used, remove the print when the developed image has the proper degree of contrast, gradation of tone, and depth of shadows. Development by inspection in this way requires some experience.

Step 5. Remove the print from the developer and allow it to drain for a few moments. Then place it in the water bath and rinse for about 15 seconds to remove the developer and prevent its being carried over into the fixing bath. Some workers use a short-stop solution, consisting of a solution of acetic acid (28%) in water (1:20), instead of the water bath to prevent any continued development or any staining of the print. Either method is acceptable.

Step 6. Transfer the print to the fresh fixing bath and leave it for 15 to 20 minutes. Agitate the solution frequently to ensure exposure of all parts of the print to the fixer and to prevent any staining of the print.

Fixing time must be accurate, since insufficient fixing will result in fading of the print image with time and excessive fixing will cause some loss of detail in the highlights.

Step 7. Wash the print in running water for at least 30 minutes to remove any chemicals embedded in the paper. The procedure is similar to that described for washing the negative. The print must be *thoroughly* washed, since it will turn brown and fade in time if it is not.

DRYING AND FERROTYPING

Drying and ferrotyping of glossy prints is usually done on highly polished chrome-plated metal sheets, the surfaces of which rarely require any prior cleaning.   If cleaning is necessary, do it with a cloth moistened in alcohol.   Then rinse the plate in cold water, allowing as much water as possible to remain on the surface.   Remove the thoroughly wet prints from the water one at a time and place them face down on the polished surface of the ferrotype plate, taking care that they do not touch or overlap.   Pass a rubber or plastic roller over the prints several times, so that the emulsion side of the prints is in perfect contact with the surface of the ferrotype plate and most of the water is squeezed out of the prints.   Remove the excess water with a cloth. Stand the plate upright and allow the prints to dry at room temperature.   When the prints are completely dry, which will be after about 30 minutes, they will slowly peel away from and drop off the plate.

Uneven glossiness of the prints may be caused by irregular contact with the plate surface, by uneven soaking before ferrotyping, or by an inadequately polished plate surface.   If the ferrotype plate is damaged or dirty, the prints may stick to it.

Print Defects

Prints that are too light may have been underexposed or underdeveloped, or the developer temperature may have been lower than that recommended by the manufacturer.   Conversely, prints that are too dark may have been overexposed or overdeveloped, or the developer temperature may have been higher than that recommended by the manufacturer.   Stains may be caused by overdevelopment of underexposed prints, weak developer, high temperature, inadequate agitation of the prints in the fixing solution, or careless immersion in the fixing bath. Foggy prints may be caused by outdated printing paper, overdevelopment, or the use of too intense a safelight in the darkroom. Prints may lack contrast because of underdevelopment, overexposure, or use of the wrong grade of printing paper.

Mounting Prints

CEMENT MOUNTING

There are a variety of adhesives (for example, rubber cement, dextrin, and starch) that may be used to mount photomicrographs. Rubber cement is the best, since the print will lie straight without buckling after the cement dries.

With a pencil, mark off on the mount an area slightly smaller than the trimmed print that is to be mounted. Spread rubber adhesive over this area and allow it to dry until it is tacky. Similarly, spread rubber adhesive over the back of the print and allow to dry. Carefully place the photomicrograph in position on the mount and press it gently into place, using a roller. Any excess cement can be easily removed with a rubber cement "pick," which is simply a small block of plastic foam or ball of dried cement.

DRY MOUNTING

Place on the back of the print a sheet of mounting tissue a little larger than the print. Touch the central portion of the mounting tissue with a hot iron. The heat will cause the tissue to stick to the print. Trim the print and the excess tissue.

With a pencil, mark off on the mount an area slightly smaller than the trimmed print. Position the print on the mount. Hold the print down with two fingers and raise one corner of the print, but not the tissue, with the index finger. Touch the exposed tissue at this corner with a heated iron held in the other hand, and then press the print to the mount. Repeat this procedure with a second corner, so that the print is held in place on the mount.

Place the print and mount, with the print upward, in a mounting press. Cover the print with a thoroughly clean metal plate and clamp tightly for about 20 seconds, keeping the press at a temperature of 140 to 160°F. Remove the mounted print from the press and allow it to cool.

If the tissue adheres to the print but not to the mount, the temperature of the press is too low. If the tissue adheres to the mount but not to the print, the temperature is too high.

GENERAL PHOTOMICROGRAPHIC PROCEDURE

The general procedure for setting up the microscope and focusing the specimen for photomicrography is described below. This procedure may vary slightly for different types of cameras, as well as with the choice of illumination or the use of an oil-immersion objective.

Step 1. Switch on the illuminator and determine whether the intensity of the light is safe for viewing. If not, use a neutral-density filter.

Step 2. Place the specimen slide on the microscope stage and bring it into focus.

Step 3. Set up the Köhler system of illumination as described on page 42.

Step 4. Move the slide so that the selected portion of the specimen is approximately in the center of the field of view. Bring the specimen into sharp focus by using the fine adjustment.

If an oil-immersion objective, or a method of illumination other than transmitted light, is to be used, the variations in this procedure described below should be made before photographing the specimen.

If a high-power oil-immersion objective is to be used, first set up the Köhler method of illumination using the dry 40X objective. Remove the slide from the stage carefully and place a drop of immersion oil on the top lens of the condenser. Replace the slide on the stage and put a drop of immersion oil on the cover glass. Carefully move the oil-immersion objective over the specimen and bring to a fine focus with the fine adjustment.

If an opaque object such as a polished metallurgical specimen is to be photomicrographed, reflected or incident light must be used with a standard metallurgical microscope provided with a vertical illuminator. The following is the procedure to be used in focusing the opaque specimen:

Step 1. Switch on the illuminator and place the opaque specimen on the microscope stage.

Step 2. Bring the specimen into focus and open the field iris diaphragm of the illuminator until it is slightly larger than the field of view.

Step 3. Remove the eyepiece and adjust the substage iris diaphragm so that the circle of light fills approximately four fifths of the back lens of the objective. Replace the eyepiece.

Step 4. Select the desired portion of the specimen and bring it into fine focus.

It should be noted that opaque specimens can be illuminated by the Köhler method when a vertical illuminator is used.

The general procedure for setting up the camera and photographing the specimen with a vertical- or horizontal-bellows camera is as follows:

Step 1. Load several plate-holders with sheet film or plates having the proper emulsion characteristics. Be careful not to scratch the film surface or let it come into contact with dust particles.

Step 2. Position and center the camera over the microscope eyepiece, securing it rigidly so that vibrations coming from any external source will be prevented. Set the camera for time exposure, which allows the shutter to be kept open.

Step 3. Switch on the microscope illuminator. It is assumed that the Köhler system of illumination has already been arranged and that the specimen has been brought into sharp focus with the microscope.

Step 4. Bring the specimen into focus on the ground-glass screen. Extend the bellows to the required distance to achieve the desired magnification. Refocus the specimen sharply on the ground-glass screen. If additional illumination is necessary for the ground-glass focusing, increase the light intensity during this step with the rheostat control on the illuminator. When exposing the film, decrease the light intensity with the rheostat control or a neutral-density filter.

Step 5. Determine the correct exposure time as described above.

Select the correct filter for the particular specimen and place it in the filter holder. Adjust the exposure time to include the filter-factor time.

Step 6. Recheck the focus of the specimen on the screen. Close the camera shutter and set it for the correct exposure time.

Step 7. Remove the ground-glass screen and insert in its place the film- or plate-holder, taking care not to move the assembly in any way so that the film will be in exactly the same plane as the ground glass. Remove the protective cover or dark shield from the film-holder and expose the film for the correct time.

Step 8. Replace the protective cover in the film-holder and remove the holder from the camera. Replace the ground-glass screen in the original position.

Step 9. Record any data that may be needed for taking future photomicrographs, such as the distance of the bellows extension, the objective and eyepiece used, the magnification achieved, the type of filters used, and the type and method of illumination.

Step 10. Process and print the film as discussed earlier in this chapter.

The procedure for photographing the specimen with a 35-mm or Polaroid camera is similar to that for the bellows camera, though simpler because there is no interchanging of ground-glass screen and film-holder. Load the camera with the correct film, focus it at infinity, and set the shutter for the correct exposure. Mount the camera over the eyepiece with the connecting collar. Focus the specimen sharply, using the observation eyepiece. Make the exposure, pressing the cable release gently so that no vibration occurs. Wind the film to the next frame and repeat the procedure.

Pointers for Good Photomicrography

1. If the specimen is to be photographed with transmitted light, be sure that it is properly prepared, of uniform thickness, properly stained, and mounted on a glass slide. If a metallurgical specimen is to be photographed, be sure that it is properly polished and etched.

2. Choose the proper type of illumination and adjust for proper intensity.

3. Use Köhler illumination whenever possible.

4. Adjust the substage iris diaphragm for the best possible image resolution. Do not manipulate the iris to reduce the intensity of the light, but instead use a neutral-density filter.

5. Use an eyepiece of the flat-field type.

6. Choose an objective that gives a substantial flat field free of aberrations and the desired initial magnification, and that has a large numerical aperture to give good resolution.

7. Select the desired portion of the specimen and project its image on the center portion of the ground-glass screen.

8. When the desired initial magnification is chosen, allow relatively long-bellows rather than short-bellows magnification, to give a flatter field.

9. Be sure that the camera and microscope are securely held, since any external vibration will cause the photomicrograph to be fuzzy, out of focus, and lacking in resolution.

10. Use an auxiliary magnifier to achieve the sharpest image when focusing the image on the ground-glass screen.

11. Determine the exposure time accurately to achieve the best exposed negative.

12. Use the correct colored filter or filters to achieve the desired contrast.

13. To avoid vibration, close the shutter or press the cable release gently and carefully. For the same reason, insert the plate-holder without moving the camera and microscope assembly.

CHAPTER 12

Physical and Chemical Microscopy

THE MICROSCOPE IS HIGHLY USEFUL in the physcial or chemical laboratory, where it enables much analytical information regarding the physical and chemical characteristics of substances to be obtained rapidly and easily from small amounts of material. In many cases, a substance can be conclusively identified by microscopical methods without further physical or chemical analysis.

Many substances, for example, can be studied and identified by the fusion method, which involves observation of their behavior under a microscope as they are heated to the point where they fuse or melt. The temperatures generally used in this method range from about -100 to about 375°C. This method is important in studying the behavior of organic compounds, crystal growth and structure, metals and alloys, and for speed and ease in identifying chemical compounds. Used in conjunction with crystallographic methods, the fusion method can identify unknown chemical compounds and determine the purity of known compounds.

Many organic compounds and alkaloidal substances can be identified under the microscope by systematic application of various reagents, which produce different reactions that aid in the identification of test samples. These sensitive reactions can be more easily observed under the microscope. Addition of a specific reagent to a microquantity of the test substance may produce a precipitate, which may be either crystalline or a colored or uncolored amorphous solid, that gives chemical information concerning the sample. Characteristic crystals, formed either directly or from the precipitate, can be analyzed microscopically to determine many optical properties and characteristics, which also aid in the identification of the test substance.

The general principles and methods of physical and chemical microscopy constitute the subject matter of this chapter.

EQUIPMENT FOR THE FUSION METHOD

Microscope

The standard compound microscope can be used for the fusion method, and the specimen can be viewed with transmitted, reflected, or polarized light.

Melting points or refractive indices of crystal materials can be more readily determined under a polarizing microscope with the polarizers crossed, so that the solid specimen appears bright against a dark background. The strain-free objectives generally used with the polarizing microscope should not be used with the hot stage (described below), however, since the heat will strain these objectives. Standard objectives should be used. If it is necessary to examine the specimen with strain-free objectives, it should first be cooled to within a few degrees of room temperature.

A magnification of 20X to 100X will usually be sufficient for most procedures in the fusion method. Objectives of 5X and 10X with longer working distances than normally used (at least 8 mm) should be used for this work. Such objectives can be kept as far as possible from the hot stage, to protect the lens system from damage. The condenser should also be either removed or racked down as far as possible to protect it from heat. If these precautions are observed, it is possible to use the hot stage on the microscope stage without damaging either the microscope or its optical parts.

Hot Stage

The Kofler hot stage (see Fig. 163), with which temperatures can be controlled within $\pm 0.5°C$, is a round insulated metal stage about 90 mm in diameter and 20 mm deep. It incorporates a nichrome heating unit controlled by a rheostat. The removable metal rim of

Fig. 163. Kofler hot stage mounted on a microscope stage and connected to a rheostat.

the stage and its glass cover form the heating chamber.   A separate glass heat baffle supplied with the hot stage is used to cover the mounted specimen, thus ensuring even distribution of the heat over the specimen.

The stage has a built-in condensing lens, obviating the standard substage condenser. A threaded post with a knurled knob holds such accessories as the fork clamp for glass slides and the sublimation block. A central opening or light well about 2 mm wide is positioned directly over the condenser to allow transmission of light through the stage. This light well is insulated from the heating system by a translucent glass plate.   Two adjustable brackets (see Fig. 164) permit attachment of the Kofler stage to the microscope stage.

The hot stage is supplied with two thermometers, one with a range of +30 to 230°C, the other with a range of +60 to 350°C. In use, one

Fig. 164.   Kofler
hot stage.

thermometer is inserted into an opening near the aperture of the hot stage, as close as possible to the specimen. Thermistors may be used instead of the thermometers to determine melting points, so that it will not be necessary to look away from the microscope to read the temperature.

The Mettler FP2 hot stage (see Fig. 165) measures temperatures

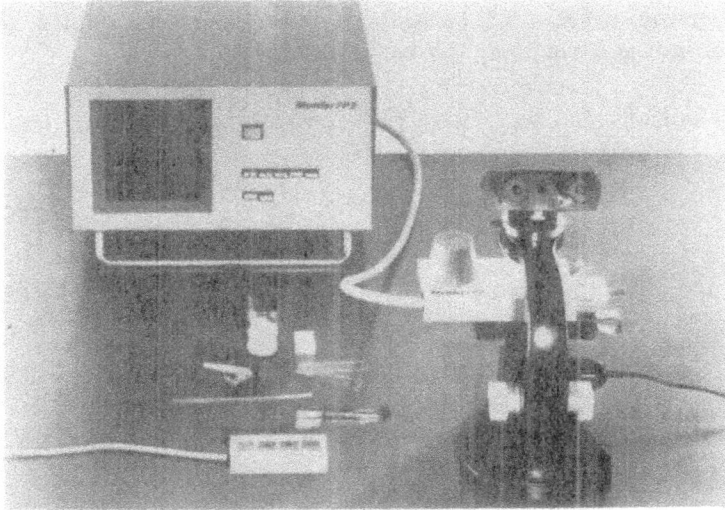

Fig. 165.   Mettler FP2 hot stage.

accurately without the use of either thermometers or thermistors. Instead, the temperatures of up to three samples under test are displayed on a digital indicator panel connected to the stage. Five pushbuttons on the panel allow the temperature to be increased or decreased at one of three constant heating rates (0.2, 2, or 10°C/minute). A remote four-pushbutton unit controls the display of the temperatures of the stage itself and the sample or samples. Most operations are easier and more precise on this stage, since the microscopist can record exact temperatures without diverting his attention from the microscope. The procedures used with the Mettler hot stage are similar to those described later in the chapter for the Kofler stage.

Hot Bar

The hot bar, or bench, is an important accessory when the fusion method is to be used. It is used to determine the approximate melting point of a test substance and to heat the sublimation block to the desired temperature before either is placed on the hot stage. Essentially, the hot bar is a stainless steel bar about 370 mm long and 38 mm wide. In use, it is heated to the maximum desired temperature at one end and the temperature decreases uniformly along the bar to the other end, where it will be at the lower point of the desired range. The heat gradient along the bar is generally linear.

The Kofler hot bench (see Fig. 166) may cover various tempera-

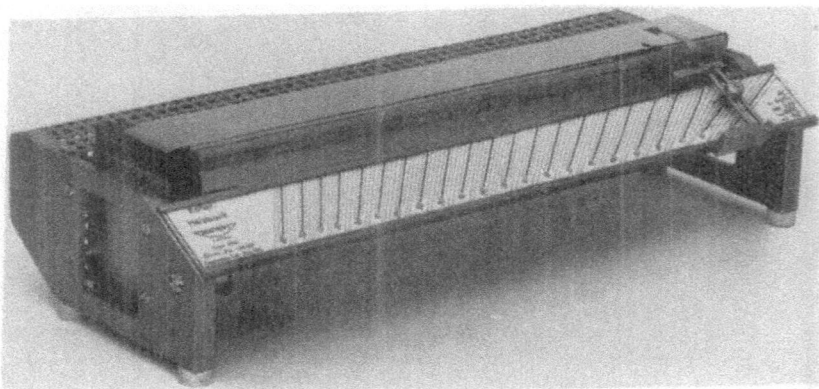

Fig. 166. Kofler hot bench.

ture ranges: +10 to 210, +50 to 260, or +80 to 180°C, the most often used range being +50 to 260°C.

Sublimation Block

A sublimation block (see Fig. 167) is used to determine the subli-

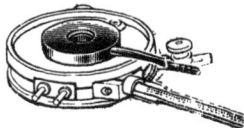

Fig. 167.   Sublimation block.

mation patterns of various compounds and, from the crystalline sublimates, the characteristics and properties of these substances.   The
block, made of a metal alloy such as chrome-plated brass, is about
50 mm in diameter.   It incorporates a slotted arm for attachment to
the hot stage, a central light well (aperture) about 15 mm in diameter,
a cemented glass bottom, and a recessed shoulder to accommodate an
18-mm round cover glass placed over the test specimen.

Cold Stage

There are several types of cold stages available.   The Thomas-Mc-
Crone cold stage (A. H. Thomas Co.) cools the sample with nitrogen
gas, which is chilled by being passed through a mixture of dry ice and
acetone in a Dewar flask and then heated as it enters the stage. Temperatures ranging from −100 to +70°C can be produced and controlled with an accuracy of ±1°C. The rate of cooling of the sample
chamber is determined by the rate at which the cooled gas enters the
stage.

The cold stage is a rectangular plastic block, $2\frac{1}{2}$ by $4\frac{3}{8}$ by 1 in.
It comprises a lower chamber for heating the cold gas, which flows in
through a gas inlet tube, and an upper chamber that holds the
sample. The sample chamber lies below the surface level of the stage
proper.   The position of the sample within this chamber is adjusted
by means of a manipulator rod. An adjustable arm attaches the cold
stage to the microscope stage.   The thermometer is protected from
breakage by a transparent plastic tube and is inserted through a gas-
keted port.

CALIBRATION OF EQUIPMENT FOR THE FUSION METHOD

The hot stage, hot bar, and cold stage must be calibrated every few months.   The procedures for calibrating these instruments are described below.

Hot Stage and Cold Stage

To calibrate the hot stage, heat it, with no accessories attached, to about 200°C to drive off any occluded moisture trapped in the instrument.   Allow the stage to cool.

Insert the thermometer into the stage.   It is important to note the position of the thermometer during calibration, since the thermometer must be in the same position for temperature measurements of test specimens if the calibration data are to be valid. Place the metal guard over the protruding portion of the thermometer to protect it and to ensure that the ambient temperature is as nearly the same as possible during calibration and subsequent tests.

To determine the heating rate, place the hot stage on the microscope stage with all the necessary accessories in position — thermometer in place, glass specimen slide on the hot stage, glass baffle over the slide, and glass cover plate over the hot stage.   Do not place the glass slide directly on the hot stage, since the slide may crack. Rather, use insulating supports to hold it.

Turn the voltage regulator or variable transformer to a low voltage reading and record the rate of increase in temperature for each minute.   Continue to record the rate until it has dropped below 2°C/minute.   At this point, stop recording the temperature, turn the current off, and allow the stage to cool. From the data and the voltage setting, determine the temperatures that correspond to heating rates of 2,...4°C,...per minute by plotting the rate as a curve of temperature against time (see Fig. 168).

Reset the voltage regulator or variable transformer to a higher voltage and repeat the procedure for determining the rate of increase in temperature for each minute.   As a rule, the procedure is repeated at voltage settings of 20, 40, 60, 80, and 100.

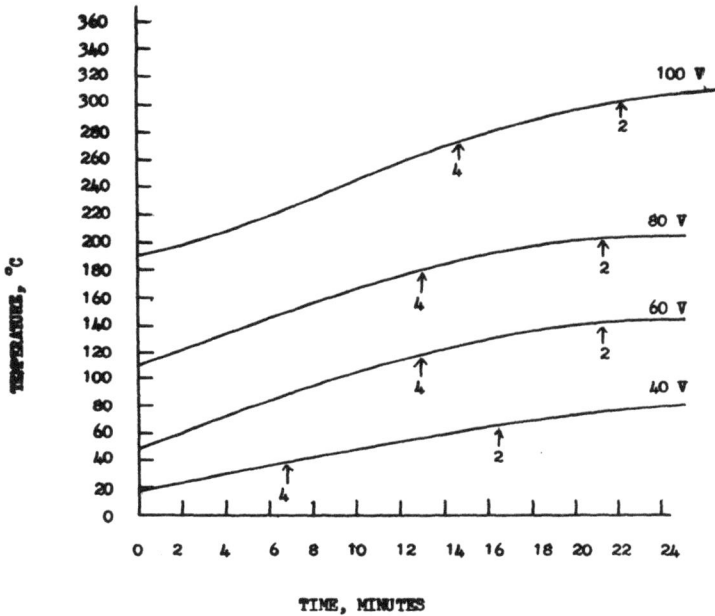

Fig. 168.   Heating-rate curves (discussed in the text): 2 and 4°C/minute.

There will now be sufficient data to determine what voltage to use to obtain a specific temperature when the rate of temperature rise is 2 and 4°C/minute. To do this, prepare another graph by plotting the voltages (or, if a voltmeter is not available, the numerical readings on the variable transformer) against the temperatures at which the rate of temperature rise is 2 and 4°C/minute (see Fig. 169). This information is important for exact determination of the melting points of substances, a procedure for which a reproducible accuracy of ±0.2°C is desirable.

The use of these data is illustrated by the following example:

An exact determination of melting point is to be made for a substance that has previously been determined to melt at approximately 94°C. A voltage is chosen from the data or graphs such that the rate of temperature rise is 4°C/minute at a temperature as close as possi-

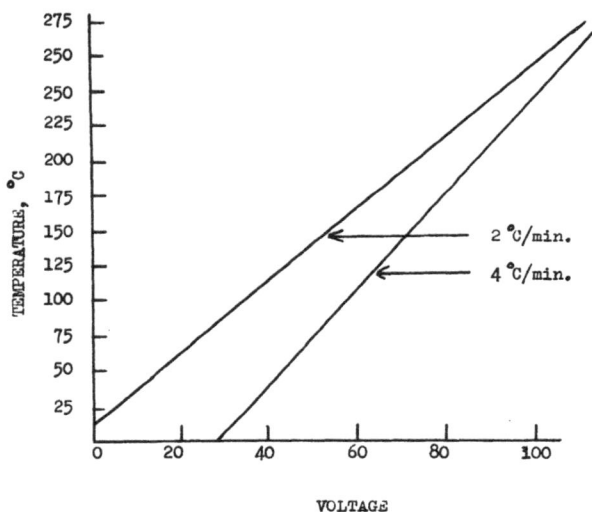

Fig. 169.   Heating-voltage curves (discussed in the text): 2 and 4° C/minute.

ble to 94°C.   As the temperature reaches about 90°C, the time is to
be checked with a stopwatch to determine the exact melting point. To
do this, watch the thermometer until the reading is exactly 93°C, and
start the stopwatch at this point. Now look again through the microscope and stop the stopwatch at the exact melting point of the test
substance — that is, when the last crystal disappears. The exact additional temperature increment can now be calculated from the elapsed
time on the stopwatch, since the rate of temperature rise is known to
be 4°C/minute.

The procedure for calibrating the cold stage is similar to that for
calibrating the hot stage, except thas it is extremely difficult to maintain a uniform heating rate with the cold stage.

Lists of standard substances suitable for calibrating the hot or cold
stage can be found in any handbook of chemistry, where exact

melting points are given for many substances. For high melting points, inorganic salts are best, since they are readily available in pure form and are stable over long periods.

Hot Bar

A hot bar should be calibrated so that approximate melting points can be more accurately determined. Generally, the bar is calibrated by using several standard compounds with successively higher melting points. A few such compounds and their melting points are given in the following list:

Compound	Melting Point (°C)
Azobenzene	68
Vanillin	81
Acetanalide	114
P-Acetophenetide	137
Sulfanilamide	164
Sulfapyridine	190
Saccharin	228
Caffeine	235–237.5

One can expect a reproducible accuracy of only $\pm 1°C$ with the hot bar, since this instrument is susceptible to temperature fluctuations in the ambient air.

To make an approximate determination of melting point with the calibrated hot bar, heat it and sprinkle a few crystals of the finely ground test specimen along it to determine the point at which the crystals melt. Sprinkle a few additional crystals over the areas where the first ones melted to locate a fine demarcation point between the melted and unmelted crystals.

Select two standard substances with melting points that bracket the approximate melting point of the test sample. Sprinkle these substances in the proper areas on the hot bar through a sieve (50–100 mesh)

The points at which the three substances melt are measured carefully on the scale attached to the hot bar, and the melting point of the test substance is then determined by interpolation. Allow at least 5 seconds for the sample to come to equilibrium before recording the melting point.

OBSERVATION OF PHYSICAL PROPERTIES BY THE FUSION METHOD

Each substance observed under the microscope may require different procedures to identify it and to determine its major characteristics, but usually the behavior during melting, the melting point, the refractive index of the melt, and the optical properties of cooled crystals will clearly identify the substance. The observations discussed below should therefore be made during the heating period on the hot stage.

Behavior During Melting

Most compounds melt and change to the liquid state when heated to a high temperature, the exceptions being compounds that sublime or decompose. Some materials decompose only slightly on heating, but give characteristic and recognizable crystals on cooling. Other compounds decompose completely when heated, however, so that no crystalline forms can be obtained when the melt is cooled.

MELTING POINT

The *melting point* of a substance is the temperature at which the solid changes to a liquid or at which the solid and liquid phases are in equilibrium under a pressure of 1 atmosphere. In practice, the temperature at which the last crystals of a sample are observed to melt under the microscope is generally taken as the melting point. If the substance is pure, it will change sharply from solid to liquid at the melting point. If impurities are present, however, the solid will not melt sharply at a definite temperature, but may melt gradually over a range of temperature before melting is complete. Generally, the melting point determined by the hot-stage technique is somewhat lower than that determined by macroscopic methods such as the capillary tube procedure.

In determining melting points, the following precautions must be observed:

The smaller the quantity of test material taken for a melting-point determination and the more finely divided it is, the greater the accuracy that can be achieved.   It has been estimated that if not more than 10% of the microscope field is covered with the test substance, an excellent melting-point determination can be made under 50X magnification.   Test samples of many substances can be finely divided by grinding the crystals between two pieces of glass.

It is important that no crystals on the glass slide adhere to the cover glass when it is applied, since any crystals on the cover glass will generally melt at about 1°C higher than those on the slide.

In calibrating the hot stage with a standard substance of known melting point, it should be remembered that the standard substance may contain a slight amount of impurities. If it does, it will not melt suddenly at the specific temperature, but gradually over a small, definite temperature range.   In such a case, the melting point is the temperature at which the last crystal disappears.

Finally, a glass baffle should always be placed over the glass slide to aid in even distribution of heat over the specimen.

Other behavior patterns of the compound can be noted at the same time as the melting characteristics are observed.   For example, the temperature at which the loss of water of crystallization occurs can be observed when the substance changes to an opaque white powder. It is therefore possible to determine by the fusion method whether certain crystals are hydrates.

SUBLIMATION

Sublimation is the transformation of a solid to a gas on heating and reversion of the gas to the solid phase on cooling. Sublimation differs from distillation in that the solid does not pass through an intermediate liquid phase.   Most organic and a few inorganic substances sublime with increasing temperature before the melting point is reached.

If a cover glass is placed over the specimen, the sublimate will usually collect on the underside of the cover glass. It is possible to determine the identity of many compounds by examining the optical properties of their sublimates. The sublimate should be transferred to a clean glass slide for examination.

The procedure for subliming an organic compound is as follows:

Place a sublimation block on the hot stage and transfer about 50 mg of the finely ground sample to the well of the block. Cover the test substance with a cover glass and gradually raise the temperature. If the temperature of the cover glass is kept close to the temperature of the block, larger and purer crystals are obtained. It is therefore desirable that the cover glass be preheated to approximately the temperature of the block before use.

POLYMORPHISM

Many substances change, when heated, from one crystalline structure to a liquid and then to a second crystalline structure before they actually melt. This property of crystallizing in two or more different systems is called *polymorphism*. A classic example of a polymorphic substance is carbon, which crystallizes as diamonds (isometric) or as graphite (hexagonal). It is possible to determine the melting point of each polymorphic form. Polymorphic behavior may also be used to identify substances.

Mixed Fusion Method

In this method, a known substance is mixed with the test sample and the mixture is fused on the hot stage. When the mixture is cooled, the crystal growth and appearance will aid in identification of the unknown substance. Identification tables (Kofler's Tables) that give analytical characteristics of about 1,200 common organic compounds with the mixed fusion method are very useful for identifying or confirming the identity of test materials.

A specific temperature range is used for examining the test substance, and two suitable reagents that fuse within this range are selected. The melting points of the test substance and the two reagents are first determined. In addition, the refractive index of the melt is determined by comparison with standard glass powders of known refractive indices.   These results serve to identify the unknown compound by reference to Kofler's Tables.

The procedure for the mixed fusion method is as follows:

After determining the approximate melting point of the test substance, place a small amount on a microscope slide and cover it with a cover glass. (The amount of test substance used should be such as to cover not more than half the area of the cover glass, after melting and cooling.)   Place a small amount of the reagent of known, lower melting point on the area of the slide adjoining the solidified test sample, outside the cover glass but in contact with it. (Some standard reagents used for the mixed fusion method are azobenzene, benzil, acetanilide, phenacetin, benzanilide, salophen, dicyandiamide, and phenolphthalein.)   Heat the slide on the hot stage until this reagent melts, begins to flow under the cover glass, and dissolves a portion of the solidified test substance.   Allow the slide to cool.   Reheat the slide gently so that all of the reagent and most of the test substance are melted.   Gradually cool the melt and observe its behavior.

This mixing procedure will have left unadulterated or pure test sample on one side of the solidified mix under the cover glass, pure standard reagent on the other side, and an area between the two pure substances in which there are zones of gradation of mixing, with the formation of crystals of varied composition.   These intermediate zones and crystal fronts can be examined microscopically during cooling, and the eutectic temperature (defined later in the chapter) can be easily determined when the mixed zone begins to crystallize rapidly. This phenomenon is quite characteristic for the test substance and is an aid in its identification.

The mixed fusion method can also be used to confirm the identity of a compound, by using a known substance that is thought to be the same as the test substance.   If the two are identical, both will flow freely into each other on heating, the resultant solid crystals will grow homogeneously at a uniform rate throughout, the crystals will appear homogeneous under the polarizing microscope, and all the interference colors will appear uniform throughout the solidified substance.   On being reheated, the solid mass will have the same melting point as the pure test specimen and the added standard reagent.   If the test substance and the added reagent are not the same, there will be a noticeable lowering of the melting point in the areas where mixing takes place.

REFRACTIVE INDEX OF THE MELT

The refractive index of the test substance, which can be an indication of its identity, can be determined with the hot stage and standard glass powders of known refractive indices. A set of 24 such standard glass powders with gradually increasing refractive indices is distributed by A. H. Thomas and Co. and by W. J. Hacker.

The method of determining the refractive index of the melt, which is discussed below, is similar to that described on page 197.

The molten test substance, with the standard particles immersed in the melt, is observed under the microscope. The Becke line is used to determine whether the refractive index of the molten test substance is greater or less than that of the glass standard. If the refractive indices of the glass and the melt are the same, the glass particles will become invisible. A red filter is generally used to eliminate any dispersion effects.

In many cases, the test substance may show more than one refractive index as the temperature of the melt increases. This phenomenon gives a quantitative characteristic of the substance.

Behavior During Cooling

The behavior of fused substances being cooled on the hot stage can be readily observed under the microscope. Many substances solidify and crystallize quickly as the temperature is lowered. Other substances do not crystallize at all as the temperature is lowered gradually, but remain in a supercooled liquid state unless induced to crystallize by seeding with a crystal or other external means.

SUPERCOOLING

When the temperature of a liquid substance is lowered below its melting point and there is no separation of solid matter, the substance is considered to be in the *supercooled* state. Many organic compounds eventually solidify, when supercooled, to an amorphous, glasslike solid. Generally, the higher the temperature to which a substance is heated above its melting point and the longer it is kept at this temperature, the greater is its tendency to remain in the supercooled state.

CRYSTAL GROWTH

When a fused substance begins to solidify during the cooling period, crystals appear to form around smaller particles of the melt called *seeds* or *nuclei*. These growing crystals will assume a characteristic *growth pattern* or *front*. Depending on the substance, this pattern or front will be one of three main types. The first pattern, which is rather rare, shows very little definite crystal form and is rather smooth. The second, more common pattern shows the formation of definite rodlike and platelike crystals, the characteristics of which can be readily used for identification of the substance. The third pattern shows the crystals growing from a number of different nuclei and forming spherical grounds of minute radiating crystals called *spherulites*. These radiating crystals are in most cases needle-shaped.

The type of crystal front formed depends on the purity of the compound, the temperature at which cooling is begun, and the polymorphic form of the substance at the moment of cooling. Usually the slower the crystal growth, the better its formation for showing definite crystal structure and behavior.

Cooling of the substance to the solid state results in a shrinkage of volume, which causes specific, characteristic behavior. The substance may show typical cracks that aid in identification. In cooling, certain molten substances may dissolve air, which appears in the solidified mass as bubbles. These bubbles, which appear in various shapes, are also very useful in determining the identity of the substance.

IDENTIFICATION OF CRYSTALS

The crystal characteristics discussed in the following paragraphs can be routinely observed when crystals are examined under the microscope and may help to identify the test substance. A magnification of 100X and transmitted light are most often used for routine initial observations. The procedures for determining the various crystal characteristics are described in Chapter 7.

Observe the crystal *structure*, uniform or varied; the *form*, crystal, geometrical, diamond- or bar-shaped; and the *contours*, regular or irregular.

Using a polarizing microscope with the polarizers crossed, note the intensity of *birefringence,* if any; the *order of interference colors*; and the main *color* of the crystal, if it is colored. Measure the *extinction angle* and the *angle of elongation*. Note whether the crystal shows *dichroism.* Rotate the stage under crossed polarizers and determine whether the crystals are *isotropic* or *anisotropic.* Examine the *interference figures,* if any, and determine the *optic sign* of the crystal. Using an eyepiece micrometer, determine the *size* of the crystals and the *ratio of length to width.*

It will also be of great value to obtain or to make *photomicrographs* of known microcrystals so that they can be compared with the test substance. Refer back to the discussion of general photomicrographic procedure in Chapter 11, page 340, if necessary.

Behavior During Reheating

Reheating the cooled melt after it has partially crystallized can also aid in determining the properties of the substance. Polymorphic transformation may occur during the reheating, and larger and better crystals may be formed when the melt is again cooled.

DETECTION OF IMPURITIES BY THE FUSION METHOD

The hot stage is highly useful in detecting possible impurities in pure compounds. The two major methods of detecting impurities are to determine the melting point and to observe how the molten substance crystallizes on cooling.

A pure substance with an added percentage of impurities is called a *eutectic* substance. When the substance is impure, the eutectic temperature is generally about 5 to 10°C lower than the melting point of the pure substance. The amount of substance, if any, that liquefies or fuses before the true melting point of the pure substance therefore indicates the approximate quantity of impurities present. If no fused droplets appear before the true melting point, it can be safely assumed that the substance is in the pure form. If a larger sample of the material is taken for the test, the amount of impurity can be determined with an accuracy of $\pm 1\%$ by this method.

Impurities can also be detected by cooling the fused substance gradually and observing its crystallization characteristics. If the impurity is anisotropic, it will be readily visible under the polarizing microscope. Moreover, the eutectic is left dispersed through the formed crystalline substance as the temperature is lowered.

PURIFICATION OF MICROSAMPLES BY THE FUSION METHOD

Small samples can be purified with the aid of the microscope and the hot stage. To accomplish this, the impure substance is placed on a glass slide, a solvent is added, and the substance is then recrystallized. The procedure for recrystallization is described below.

Since the solvent used in this procedure must not spread erratically over the slide, a solvent with a high surface tension is usually selected. Suitable solvents are water, acids, bases, aldehydes, ketones, and higher alcohols. Alternatively, the solvent can be confined by using slides made of material other than glass, such as Teflon, polyethylene, or other plastic, or a glass slide can be coated with a material that is not easily wetted by the added solvent.

To purify a sample, place about 1 mg of the substance on a slide. Add 1 drop of the chosen solvent to dissolve the sample. Place the slide on the hot stage and aid solution by warming gently. It is essential that the solvent dissolve sufficient sample to make a saturated solution.

Cool the solution slowly. The first crystals will appear near the edge of the liquid. Push these crystals back into the center of the liquid with a pointed glass rod. Gradually, well-formed crystals will appear throughout the liquid. Separate the pure crystals from the liquid by pushing them along the slide with the rod until they are on a dry area.

Another method of isolating the crystals is to touch the liquid surrounding them with a piece of filter paper moistened with the solvent. If desired, the liquid can then be recovered from the filter paper with the use of an appropriate solvent and the crystals can be recrystallized again for greater purity.

CHEMICAL MICROSCOPY WITH REAGENTS

As noted at the beginning of this chapter, many organic and inorganic compounds can be identified under the microscope by systematic application of various reagents and study of the resulting reactions and solid precipitates, crystalline precipitates, or crystallized solid precipitates. The techniques for microchemical analysis and identification of such substances, as well as other related substances, are discussed in the following text.

Identification of Water-Formed Deposits

Water-formed deposits include any insoluble materials that have been formed by the reaction of water with any material in contact with it. They include scale deposits, sedimentary sludge, corrosion products, and deposits of decayed biological organisms.

In carrying out the procedures for analyzing these materials, a length of fine platinum wire (0.40 mm, 26 gauge) with a tiny loop about 0.75 mm in diameter formed at one end and bent at a slight angle to the straight wire is an excellent instrument for applying the necessary reagents to the specimen. The loop can be formed by winding the end of the wire around a needle.

To prepare a sample, dry a small portion of the substance at 105°C. Crush but do not grind the dried sample in a mortar in order to obtain small particles, rather than powdery material.

COMPONENTS SOLUBLE IN WATER

Place a pinch of the crushed sample on a slide. Add a drop of water to the sample and warm gently over a microburner. Allow the slide to dry at room temperature and place it on the microscope stage. Under 50X magnification (5X eyepiece, 10X objective), observe whether any crystallization has taken place near the drop of water. If so, the sample contains one or more water-soluble components.

COMPONENTS SOLUBLE IN ORGANIC SOLVENTS

Place a pinch of the crushed sample on a glass slide and add a drop of carbon tetrachloride. Allow the sample to dry at room temperature. Examine under the microscope to determine whether there is any residue at the point where the organic solvent was added. If there is, organic components may be present in the test sample.

COMPONENTS SHOWING EFFERVESCENCE WITH ACID

Place a pinch of the crushed sample on a glass slide. Dip a glass rod in a dilute (1:4) solution of hydrochloric acid and allow the acid to flow from the rod onto the test substance. Observe whether gas is evolved, as evidenced by effervescence, or bubbling. If so, carbonates, sulfides, or metals that cause the evolution of hydrogen gas may be present. Allow the liquid to evaporate on the slide and observe whether any crystals form. Cubic crystals may be sodium chloride formed by reaction with the acid.

ALKALINE OR ACIDIC COMPONENTS

Place a pinch of the crushed sample on a glass slide and add a drop of water. Place the slide on the microscope stage and then add a drop of phenolphthalein solution. Observe whether the solution turns pink around any component in the substance. This pink color indicates that the component is alkaline.

Add a drop of water to another pinch of the crushed test sample placed on a slide. Place the slide on the microscope stage and add a drop of methyl orange solution. If any acidic component is present, the solution surrounding that component will turn red.

METALLIC ELEMENTS

Place a pinch of the crushed sample on a glass slide and add a drop of silver nitrate (5%, acidified with dilute nitric acid). Examine the slide under low-power magnification. Any matallic elements present will replace the silver in the silver nitrate, and crystals of metallic silver will be observed.

CALCIUM

Add a drop of dilute sulfuric acid (1:20) to a pinch of the crushed sample placed on a glass slide. If calcium is present, it will form crystals of calcium sulfate in the form of radiating needles.

MANGANESE

Place a pinch of the crushed sample on a glass slide, add 2 drops of concentrated hydrochloric acid, and evaporate to dryness over a microburner. Add 2 drops of dilute nitric acid (1:19) to the residue and warm over the microburner. Allow to cool. Add a small pinch of powdered sodium bismuthate to the liquid and stir with a glass rod. If manganese is present, a magenta color will be observed under the microscope.

PREPARING A TEST SOLUTION

In carrying out the remainder of the tests described below, it will be necessary to use the following test solution:

Test Solution

Crushed sample	about 50 mg
Nitric acid (conc.)	5 drops
Distilled water	1 ml

To prepare the solution, place the crushed sample in a porcelain dish. Add the concentrated nitric acid and heat gently in a hood, using a microburner, until the solution is evaporated to dryness. Add the distilled water to the dried material and stir. Allow the solid material to settle and use the supernatant liquid.

SODIUM

Place a drop of the test solution on a glass slide and evaporate to dryness. Add a drop of water to the dried residue and draw the liquid to a clean area of the slide with the platinum wire. Add a drop of zinc uranyl acetate solution (5%). Colorless or pale yellow monoclinic crystals, clearly visible under the microscope, are formed if sodium is present.

POTASSIUM

Place a drop of the test solution on a glass slide and evaporate to dryness. Add a drop of water to the dried residue and draw the liquid to a clean area of the slide with the platinum wire. Add a drop of chloroplatinic acid solution (15%, acidified with hydrochloric acid), and view the slide under 100X magnification. If potassium is present and ammonia is absent, octahedral or, in some cases, hexagonal crystals are seen. If ammonia or ammonium radical is known to be present, ignite the test sample before applying the reagent.

NICKEL

Place 2 drops of the test solution on a glass slide.   Add a drop of concentrated ammonium hydroxide solution to the test liquid and warm gently over a microburner. Insert a thin platinum wire or glass rod into the liquid and draw some of the solution onto a clean area of the slide. Evaporate this portion to dryness, and add a drop of ammonium hydroxide and a drop of dimethlyglyoxime solution (powder dissolved in a little water).   Stir the solution with a thin glass rod and observe under high-power magnification. If nickel is present, pink or magenta fine needle-shaped crystals will be seen.

LEAD

Place a drop of the test solution on a glass slide.   Add a crystal of potassium iodide to the drop. Place the slide on the microscope stage under 100X magnification and note whether there is a reaction.   If lead is present, yellow hexagonal plates and thin scales will be seen. The color of the scales may sometimes appear greenish or brownish.

Other organic or inorganic components can be identified by conducting many other tests with different reagents.   The subjects of qualitative and quantitative organic and inorganic microchemical analysis are large ones, and full discussion of them is far beyond the scope of this book.   Standard reference texts on these subjects may be consulted by the student interested in conducting further tests.

MICROCHEMICAL COUNTING AND MEASURING TECHNIQUES

Other microanalytical methods are based on counting and measuring the various ingredients in the field of view.   Several such procedures are described below, in order to demonstrate how the microscope has proven invaluable in the qualitative and quantitative determination of the components of materials.

Components of a Mixture

The microscope is especially useful for studying mixtures of powdered materials. In many cases, the individual components can be separated and counted. If carefully done, this separating and counting technique will often yield a reasonably accurate determination of the composition of a mixture. When this procedure is undertaken, note must be made as to whether the various components of the mixture are of uniform size, as well as whether they are evenly distributed in the mixture and in the microscope field of view, so that they are representative of the entire mixture. If the particles are not uniform (as might be true, for example, of different fibers in a paper sample), the counts of the particles of the various components are often weighted according to their relative volumes. In many cases, it may be necessary to count many fields and average the counts.

Separation of the components can be done with a fine-pointed forceps, a fine-pointed medicine dropper, or a fine, stiff wire. Moistening the wire with a little glycerin sometimes facilitates separation.

If one or more of the components are insoluble in specific solvents, it is helpful to add a liquid with a refractive index equal to that of one of these components. Such a liquid will render the particles of that component almost invisible, so that particles of the other components are better defined and thus easier to count.

Fiber Content of Papers

In an industrial laboratory, the fiber content of papers is identified and determined quantitatively by separating and counting the various fibers under the microscope. Various stains are used to aid in this determination. The most widely used stain is Hertzberg's stain, which imparts distinctive colors to the more important paper fibers. The preparation of Hertzberg's stain is described in the Appendix, page 405. The technique in which this stain is used is discussed below.

Place a very small piece of the paper sample in a test tube and add about 5 ml of 2% sodium hydroxide solution. Heat the tube gently to loosen the fibers in the paper. Allow to cool, then rinse the sample with several washings of water to rid it of any excess sodium hydroxide.   Add a few milliliters of dilute (1:3) hydrochloric acid to neutralize sodium hydroxide.   Wash the paper sample several times by adding water to the test tube, shaking gently, and decanting the water.

Remove the sample from the test tube and roll and press it between the fingers to further loosen the fibers.   Replace the sample in the test tube, add water, and shake vigorously until it has distintegrated into minute fibers.

Place a few milliliters of the fiber suspension on a microscope slide and allow to dry in air or in an oven.   Apply a few drops of Hertzberg's stain to the fibers and cover with a cover glass.   Examine the slide under 100X magnification with a microscope equipped with an eyepiece micrometer.

Count the number of fibers of each color seen in the microscope field.   A good technique is to count five or more fields for each component fiber in order to obtain a more accurate, average count.

Since the various fibers are of different lengths and diameters, a size factor is usually used for each type of fiber in order to determine the quantities of each fiber more accurately.   The count of fibers of each color is multiplied by the appropriate factor to obtain the actual percentage of the fiber present.

The following list shows the Hertzberg's stain colors and the size factors for three types of paper fibers:

Type of Fiber	Stained Color	Size Factor
Groundwood pulp	intense yellow	1.30
Rag (linen and cotton)	purplish pink	1.00
Sulfite and Kraft wood pulp	dark blue	about 1.2

Fiber Content of Textiles

The microscope is invaluable in the identification and quantitative determination of textile fibers.   Photomicrographs of various fibers (longitudinal and cross-sectional views) are also useful for comparison purposes when identifying a fiber. Photomicrographs of many natural and artificial fibers can be obtained from the American Association of Textile Chemists and Colorists. Pure samples of the fibers for comparison under the microscope are another invaluable aid.

It may be necessary to boil the specimen in water to remove starches and other foreign materials. If the fibers are not already separated, they should be teased apart and broken up into fiber cells with a sharp needle.

The general procedure for examining a textile or fiber specimen under the microscope is as follows:

Place the specimen on a glass slide and tease the threads apart with a sharp-pointed needle, as mentioned above. Cover the specimen with a cover glass or another glass slide. Mount the fibers in glycerin to aid in observation. Place the slide on the microscope stage and observe the fiber characteristics enumerated and discussed in the following paragraphs.

FIBERS WITH SURFACE SCALES

Fibers with surface scales include animal hairs such as wool, mohair, camel hair, cashmere, horsehair, llama, vicuña, and alpaca. These various fibers can be differentiated by observing the size and shape of the scales. Observe whether a medulla is present, the diameter of the fiber, the length of the scales (in micrometers), any color pigments present in the cortex, and whether color, if present, is diffuse or granular.

View the cross section of the fiber to determine the contour of the fiber and medulla (oval, round, or kidney-shaped) and the distribution of any pigments.

TWISTED OR BENT FIBERS

Fibers with twists or bends include cotton, kapok, and tussah silk. Cotton can be easily identified: It is a broad fiber, with frequent twists, and it has a broad lumen. Mercerized cotton has no twists, is rounder, and has either no lumen or a very narrow one. Application of Hertzberg's stain to a cotton fiber colors it reddish-violet. Tussah silk fibers also show longitudinal twists or striations and are broad,

with particles of colored pigment. Parallel or diagonal lines can sometimes be seen along the fiber. Kapok fibers are hollow circular tubes,
with many bends but no twists. Kapok has a broad lumen, but thinner walls than cotton.

FIBERS WITH JOINTS OR SWELLINGS

Fibers with joints or swellings include the vegetable stem fibers
(flax, hemp, jute, and ramie), the distinguishing characteristics of
which are: Jute does not appear jointed. Flax has pointed cell ends.
In cross section, flax has a round lumen; hemp and ramie have irregular lumens.   If reference standards are available, it will not be difficult to identify these fibers, since they have a distinctive longitudinal
appearance.   Treating the fibers with Hertzberg's stain will color flax
brownish-violet, hemp and ramie violet, and jute yellow orange.

FIBERS WITH NO DEFINITE MARKINGS

Fibers with no definite markings include the majority of the artificial fibers.   They do not have a typical cross-sectional appearance,
and in many cases specimens may differ in appearance from standard
photomicrographs because of different methods of manufacture.
Generally, physical and chemical tests must be made in conjunction
with microscopical tests to confirm the identification of these fibers.
Some of the fibers included in this group are nylon, acrilon, acetate,
asbestos, arnel, dacron, dynel, glass, orlon, polyethylene, rayon, vin-
yon, vicara, darvan, and vycron.

PHYSICAL CHARACTERISTICS

Physical characteristics such as the refractive index, birefringence
(numerical value), and melting point are also used to identify these
synthetic fibers. The refractive indices and birefringence values can be
found in the literature.   The melting points of synthetic fibers can be
determined by using a hot stage.

The procedure for determining the melting or softening points
of synthetic fibers is somewhat different from that for determining
the melting points of crystalline substances.   The procedure is as
follows:

Determine the approximate melting point of the test fiber on a hot
bar.   Place 3 mg of the test specimen on a cover glass and cover it
with another glass.   Place in position on the hot stage and raise the

temperature rapidly until it is about 20°C below the expected melting point. Press on the upper cover glass lightly with a spatula. Now raise the temperature at a rate of 3°C/minute. Note the melting point, which is the temperature to the nearest degree at which the specimen is observed to flow.

The melting points of several synthetic fibers are as follows:

Fiber	Melting Point (°C)
Acetate	260 ± 0.5
Dynel 97	185 ± 5.0
Dacron	256 ± 1.0
Nylon 6	215 ± 0.5
Nylon 6–6	254 ± 1.0
Polyethylene (high-density)	180 ± 1.0
Saran	164 ± 0.5

Fixatives, Stains, and Other Reagents for Microscopy

HISTOLOGICAL FIXATIVES

Allen's Fluid

Histological Fixative

Picric acid (with 10% water)	1.1 g
Chromic acid	1.0 g
Urea	1.0 g
Glacial acetic acid	10 ml
Formaldehyde (40%)	15 ml
Water	75 ml

Used for fixing meiotic figures in small histological specimens. Fixing is generally complete in 6 to 24 hr. For optimum results, use alcoholic staining solutions.

Bouin's Fluid

Histological Fixative

Picric acid (saturated aqueous solution, saturation point at 20°C, 1.2 g per 100 ml of water)	75 ml
Formaldehyde (40%)	25 ml
Glacial acetic acid	5 ml

A good general fixative for botanical specimens and animal tissues, especially recommended for beginners. It penetrates rapidly without undue hardening of the tissues.

Do not allow the tissues to remain in the fixative for long periods of time, since difficulty in cutting the specimens will result. Difficulty may then also be encountered in staining the nuclei of the cells. It will, moreover, be difficult to wash the picric acid from the tissues. Traces of this chemical may interfere with staining.

Fixing is generally complete in 6 to 24 hr. Fixative does not interfere with staining.

Carnoy's Fluid

Histological Fixative

Chloroform	30 ml
Absolute alcohol	60 ml
Glacial acetic acid	10 ml

A general cytological fixative. It penetrates the tissues rapidly, dehydrates the tissues, and improves nuclear staining; however, the fluid causes excessive shrinkage and destroys red cells.

Fixing is generally complete in 1 to 2 hr. Fixative does not interfere with staining.

Carnoy and Lebrun's Fluid

Histological Fixative

Mercuric chloride	(to saturation) 25 g
Chloroform	33 ml
Absolute alcohol	33 ml
Glacial acetic acid	33 ml

A fixative used for penetration of hard-shelled organisms. It is not used with specimens containing fat.

Fixing may take from 10 min to 1 hr. Fixative does not interfere with staining.

FAA Fluid

Killing, Fixing, and Preserving Mixture

Ethyl alcohol (95%)	50 ml
Glacial acetic acid	5 ml
Formaldehyde (40%)	10 ml
Water	35 ml

A popular type of killing, fixing, and preserving liquid. The fluid is stable, has excellent hardening characteristics, and the specimen may be stored for years in this liquid. Not recommended, however, for fine cytological work.

Fixing is generally complete in 1 to 2 days. Fluid does not interfere with staining.

Farmer's Fluid

Histological Fixative

Anhydrous ethyl alcohol	75 ml
Glacial acetic acid	25 ml

Fluid for killing protoplasm by rapid and violent dehydration. It is used to fix structures that are rather impermeable or for specimens that must be prepared hurriedly. Note, however, that this fluid causes excessive shrinkage of large tissues and destroys red cells.

Fixing is generally complete in 30 to 60 min. Fixative does not interfere with staining.

Formol Saline Solution

Histological Fixative

Sodium chloride	0.9 g
Formaldehyde (40%)	10.0 ml
Water	90.0 ml

The solution is an all-purpose useful fixative and preservative for micro-anatomical specimens. It penetrates well and allows for a large variety of staining methods.

The fixative is perishable in time because the formaldehyde is oxidized to formic acid, which inhibits staining. It should therefore be prepared fresh approximately every 4 months.

Fixing of small specimens is generally complete within 24 hr. Large specimens may require several weeks.

Gate's Fluid

Histological Fixative

Chromic acid	0.7 g
Glacial acetic acid	0.5 ml
Water	100 ml

Similar in action to Lo Bianco's fluid, this fixative is weaker in concentration and therefore generally preferred by botanical workers.

Gilson's Fluid

Histological Fixative

Mercuric chloride	2.0 g
Nitric acid (conc)	1.8 ml
Glacial acetic acid	0.4 ml
Ethyl alcohol (95%)	10.0 ml
Water	88.0 ml

An excellent general fixative for histological and cytological specimens. Glass or plastic tongs must be used when handling specimens.

Fixing is generally complete in 15 to 30 min for small specimens, 1 to 2 hr for medium specimens, and about 6 hr for large, dense specimens. Fixative does not interfere with any preferred staining procedure.

Heidenhain's Fluid

Histological Fixative

Potassium dichromate	1.8 g
Mercuric chloride	4.5 g
Glacial acetic acid	4.5 ml
Formaldehyde (40%)	10.0 ml
Water	90.0 ml

Like Helly's fluid, prepared prior to use; fixation and washing of the specimens are done in the dark.

The fixative can also be prepared as two solutions so that it is less perishable; one solution contains the acetic acid and formaldehyde, the other solution contains the remaining ingredients.

Fixing may take about 24 hr.

Helly's Fluid

Histological Fixative

Potassium dichromate	2.5 g
Mercuric chloride	5.0 g
Formaldehyde (40%)	10.0 ml
Sodium sulfate	1.0 g
Water	90.0 ml

Similar to Zenker's formula, except that formaldehyde replaces the acetic acid.

This solution is prepared prior to use. Fixation and washing of the specimens are done in the dark. Light hastens the reduction of the potassium dichromate by the formaldehyde.

Fixing requires about 24 hr.

Kolmer's Fluid

Histological Fixative

Potassium dichromate	1.8 g
Uranyl acetate	0.75 g
Formaldehyde (40%)	3.6 ml
Glacial acetic acid	9.0 ml
Trichloroacetic acid	4.8 ml
Water	87.0 ml

An excellent fixative for nerve structures. It is also used as a general-purpose fixative. The specimen is generally fixed overnight and then washed thoroughly in running water.

Lavdowsky's Fluid

Histological Fixative

Potassium dichromate	5.0 g
Mercuric chloride	0.15 g
Glacial acetic acid	2.0 ml
Water	100 ml

The acetic acid is added immediately before use.

A formula similar to Zenker's, except that less mercuric chloride is used. This permits longer periods of fixation without the danger of excessive hardening of the specimens. Widely used as a fixative for botanical specimens.

Fixation requires from 12 to 36 hr, depending on size and density of the specimen. Fixative does not interfere with staining.

Lo Bianco's Fluid

Histological Fixative

Chromic acid	1 g
Glacial acetic acid	5 ml
Water	100 ml

Suitable for small invertebrates and marine forms. The specimen is left in the fixative for 30 min for small forms to 24 hr for larger specimens. The specimen is then washed in water until the wash water is free of color.

Navashin's Fluid

Histological Fixative

Chromic acid (1%)	75 ml
Glacial acetic acid	5 ml
Formaldehyde (37–40%)	20 ml

Prepare the fluid immediately prior to use as two solutions by keeping the chromic acid mixture separate from the formaldehyde.

An excellent fixing, hardening, and preserving agent, especially for plants.

Fixation may take from 6 to 12 hr. Fixative does not interfere with staining.

Orth's Fluid (Stock Solution)

Histological Fixative

Potassium dichromate	12 g
Sodium sulfate	1 g
Distilled water	900 ml

Prior to use, add 1 part by volume of formaldehyde solution (40%) to 9 parts of the stock solution.

An excellent all-purpose fixative especially for nerve tissues. It is often used instead of Zenker's fluid, since it does not contain mercuric chloride and therefore does not require the special iodine rinsing.

Fixation of the specimen takes 12 to 24 hr, depending on the size of the specimen. Fixative does not interfere with staining.

Schaudinn's Fixing Fluid

Fixative for Fecal Smears

Mercuric chloride (saturated: 8.0 g per 100 ml)	65 ml
Glacial acetic acid	5 ml
Ethyl alcohol (95%)	35 ml

A fluid used for fixing protozoa, parasitic organisms, and fecal smears.

Fixing may take from 15 to 30 min. The fixative is generally warmed to 40°C prior to use. An iron hematoxylin stain is preferred for staining.

Tellyesnicky's Fluid

Embryological Fixative

Potassium dichromate	3 g
Glacial acetic acid	5 ml
Distilled water	100 ml

Dissolve the potassium dichromate in the water. Add the acetic acid just before use.

Used as a general histological fixative and in embryology.

Fixing is complete in from 1 to 2 days. The fixative does not interfere with any chosen staining procedure, although hematoxylin stains are preferred.

Zenker's Fluid

Histological Fixative

Mercuric chloride	5.0 g
Potassium dichromate	2.5 g
Glacial acetic acid	5.0 ml
Sodium sulfate	1.0 g
Water	100 ml

Dissolve the mercuric chloride and potassium dichromate in the water. Add the acetic acid just before use.

A widely used, all-purpose microanatomical and cytological fixative, excellent for protein. It does not interfere in the later staining procedures. Glass or plastic forceps, not metal, are required for handling the specimens.

Fixative of the specimen requires from 12 to 36 hr, depending on size and density. Fixative does not interfere with staining.

HISTOLOGICAL STAINING SOLUTIONS

Albert Staining Solution (Laybourn Modification)

Stain for Diphtheria Bacillus

SOLUTION A

Toluidin blue O	0.15 g
Malachite green or methyl green	0.20 g
Ethyl alcohol (95%)	2 ml
Glacial acetic acid	1 ml
Distilled water	100 ml

Dissolve the toluidin and malachite green in the alcohol and dilute with the water.   Add the acetic acid, allow to stand overnight, then filter.

SOLUTION B

Iodine crystals	2 g
Potassium iodide	3 g
Distilled water	300 ml

Mix the ingredients in the order given.

The solutions are used separately during the staining procedure.   The granules of the diphtheria bacilli stain black, the bars dark green, the intermediate portions light green.

Auerbach's Methyl Green-Fuchsin Staining Solution
Histological Stain

SOLUTION A

Methyl green	1 g
Water	100 ml

SOLUTION B

Acid fuchsin	1 g
Water	100 ml

The section is first immersed in Solution A, washed, and then immersed in Solution B.   The staining solutions are used primarily to show mitotic figures.

Best's Carmine Staining Solution
Histological Stain

STOCK SOLUTION

Carmine	2 g
Potassium carbonate	1 g
Potassium chloride	5 g
Ammonium hydroxide, concentrated	20 ml
Distilled water	60 ml

Dissolve the carmine, potassium carbonate, and potassium chloride in the water.   Aid the solution by boiling gently for several minutes until the color of the solution darkens.   Allow to cool and filter.   Add the ammonium hydroxide and allow to ripen for 24 hr.   Store in a refrigerator.   The stock solution may be kept for several weeks.

SOLUTION FOR USE

Stock solution	20 ml
Methyl alcohol	30 ml
Ammonium hydroxide, concentrated	30 ml

Used as a stain for glycogen.   Stains glycogen pinkish red.

Biebrich's Scarlet Staining Solution

Histological Stain

Biebrich scarlet (1%)	90 ml
Acid, fuchsin (1%)	10 ml
Glacial acetic acid	1 ml

A very good cytoplasmic stain since it does not overstain.

Blueing Solution

For Staining Procedure

Sodium bicarbonate	0.1 g
Water	100 ml

Used to intensify the dye during staining procedure.

Carmine Alum Staining Solution

Histological Stain

Ammonium alum	5 g
Carmine	2 g
Thymol	1 g
Distilled water	100 ml

Add the alum and carmine to the water and boil for 1 hr. Allow to cool and add enough water to make up to 100 ml. Add the thymol to prevent mold growth.

Stain for celloidin sections. The section is placed in the stain and kept at 60°C for 2 to 3 hr.

Delafield's Alum Hematoxylin Staining Solution

Histological Stain

Hematoxylin	1 g
Ammonium alum	3 g
Glycerin	10 ml
Methanol	10 ml
Ethyl alcohol (abs)	4 ml
Water	80 ml

Dissolve the hematoxylin in the methyl and ethyl alcohols. Separately, dissolve the alum in the water. Then mix the two solutions. Filter the solution and add the glycerin. Allow to ripen for a few months by exposing the loosely stoppered bottle to light.

Used for routine staining in botany and zoology. Any counterstain may be subsequently used.

Differentiating Solution

For Staining Procedure

Ethyl alcohol (70%)	100 ml
Hydrochloric acid (10%)	1 ml

Dobell's Hematoxylin Staining Solution

Histological Stain

MORDANT SOLUTION
Iron alum	1 g
Ethyl alcohol (70%)	100 ml

STAINING SOLUTION
Hematein	1 g
Ethyl alcohol (70%)	100 ml

DIFFERENTIATING SOLUTION
Hydrochloric acid, concentrated	1 drop
Ethyl alcohol (70%)	100 ml

No ripening period is required for the staining solution.

Gives intense colorations to specimen, used widely for protozoan smears. Transfer the specimens to the mordant solution for about 10 min and then, without rinsing, transfer them to the staining solution and allow to remain for about 10 min. Rinse rapidly in ethyl alcohol (70%) and transfer to the differentiation solution.

Ehrlich's Acid Alum Hematoxylin Solution

Histological Stain

Hematoxylin	2 g
Ethyl or methyl alcohol (95%)	100 ml
Glacial acetic acid	10 ml
Glycerin	100 ml
Potassium alum	10 g
Water	100 ml

Dissolve the hematoxylin in the alcohol, add the acid and then the glycerin.

Dissolve the alum separately in hot water and pour the warm mixture slowly into the hematoxylin solution with constant stirring. Store in a bottle and allow to stand uncorked, exposed to the sun and air, for about 2 months. Shake the solution thoroughly once a week. The solution is ripened when it turns a deep red. After ripening, cork the bottle tightly. The stain solution keeps well for many years. The solution can be ripened immediately by adding 0.4 g of sodium iodate to the solution and then shaking. No differentiation is necessary when using this stain. Used for routine staining in botany and zoology. It gives a reddish stain instead of blue and does not overstain.

Eosin Contrast Staining Solution

Histological Contrast Stain

Ethyl eosin or eosin y	0.5 g
Ethyl alcohol (95%)	25 ml
Distilled water	75 ml

Add a few small crystals of thymol to prevent the growth of molds. If growths occur, filter the eosin solution prior to use. The solution will keep indefinitely.

Extensively used as a counterstain after hematoxylin staining.

Feulgen Sulfurous Acid Solution

Rinse used with Schiff's Reagent

Potassium metabisulfite	1 g
Hydrochloric acid	10 ml
Distilled water	200 ml

Dissolve the potassium metabisulfite in the water and hydrochloric acid and filter.

Used as a rinse after staining the specimen with Schiff's reagent.

Giemsa's Stain (Stock Solution)

Histological Stain

Giemsa stain	0.6 g
Glycerol	50 ml
Methyl alcohol	50 ml

Transfer the Giemsa stain to glycerol and place in an oven kept at 55 to 60°C for about 2 hr. Heat the methyl alcohol on a water bath or oven to 60°C. Add to the stain solution. Allow to cool. Filter and store in a stoppered bottle.

The staining solution is made up by diluting 1 vol of Giemsa's stock stain solution with 50 vol of buffered water (pH 7).

An excellent stain for sections, films, and smears. Mount specimen in cedarwood oil or clarite, since the stain will fade with a balsam mount.

Gram Stain (Hucker Modification)

General Differential Bacteriologic Stain

CRYSTAL VIOLET SOLUTION

Crystal violet	4 g
Ethyl alcohol	20 ml
Ammonium oxalate	0.8 g
Distilled water	80 ml

Dissolve the crystal violet in the alcohol. Dissolve, separately, the ammonium oxalate in the water. Dilute the crystal violet solution by adding distilled water (1 to 10). Mix 1 vol of this crystal violet solution with 4 vol of the ammonium oxalate solution.

GRAM'S IODINE SOLUTION

Iodine	1 g
Potassium iodide	2 g
Sodium bicarbonate solution (5%)	60 ml
Distilled water	240 ml

Dissolve the iodine and potassium iodide in 5 ml of the water and then dilute with the remaining water. Add the sodium bicarbonate solution. Prepare a fresh iodine solution when a loss of color becomes apparent.

COUNTERSTAIN SOLUTION

Safranin	0.25 g
Ethyl alcohol	10 ml
Distilled water	100 ml

Dissolve the safranin in the alcohol and then dilute with the water.

Used to determine the gram-staining properties of bacteria.

Gray's Contrast Staining Solution

Histological Double-Contrast Stain

Ponceau 2 R	0.4 g
Orange II	0.6 g
Water	100 ml

Stains different tissues in different shades of orange.

Grenacher's Alcoholic Borax Carmine Solution

Nuclear Stain

Borax	4 g
Carmine	3 g
Ethyl alcohol (70%)	100 ml
Water	100 ml

Add the borax and carmine to the water and boil for 30 min. Cool the solution and add the alcohol. Allow to stand several days, then filter.

Generally used as indirect stains for whole mounts, but not for section staining. For direct staining, the solution must be diluted extensively with an alum solution.

Harris's Hematoxylin Staining Solution

Histological Stain

Hematoxylin	1 g
Absolute alcohol	10 ml
Potassium alum	20 g
Mercuric oxide	0.5 g
Distilled water	200 ml

Dissolve the hematoxylin in the alcohol. Dissolve the alum separately in the water with the aid of heat and then mix the two solutions together. Bring this mixture to a boil and add the mercuric oxide. As soon as the solution turns dark purple, remove from the flame and cool immediately by plunging the container into cold water. Potassium permanganate (0.177 g) may be added in the cold if the mercuric oxide is not available. The addition of acetic acid (4%) may improve the staining of cell nuclei.

Heidenhain's Iron Hematoxylin Staining Solution

Cytological Stain

STOCK SOLUTION
Dissolve 10 g of hematoxylin in 100 ml of absolute alcohol and allow to
ripen for about 5 to 6 weeks by exposing to air and light.

SOLUTION A

Hematoxylin stock solution	5 ml
Ethyl alcohol (95%)	5 ml
Distilled water	9 ml

Mix and filter before use.

SOLUTION B

Ferric ammonium sulfate	3 g
Water	100 ml

The solutions are used separately. Solution B is used to differentiate the
stain.

Excellent as cytological stains. Mitochondria and mitotic figures are stained
black.

Hitchcock and Ehrlich Staining Solution

Stain for Plasma Cells

SOLUTION A

Malachite green	0.3 g
Distilled water	15 ml

SOLUTION B

Acridine red	0.9 g
Distilled water	45 ml

Mix the two solutions and then filter.

Nuclei of plasma cells are stained blue green and cytoplasm crimson. Other
cells are stained in lighter shades.

Johansen's Safranin Staining Solution

Nuclear Stain

Safranin	1 g
Sodium acetate	1 g
Formaldehyde (40%)	2 ml
Ethyl alcohol (95%)	25 ml
Methyl cellosolve	50 ml
Water	25 ml

Dissolve the dye in the methyl cellosolve, and add the alcohol. Dissolve separately the sodium acetate in the water and formaldehyde and then combine with the dye solution.

Kornhauser's Hematein Staining Solution

Histological Stain

Hematein	0.5 g
Potassium aluminum sulfate (saturated)	500 ml
Ethyl alcohol (95%)	10 ml

Grind together the hematein and the alcohol in a glass mortar. Add the saturated potassium alum solution to the hematein tincture. The solution does not have to be ripened and is ready for use. The stain has good keeping characteristics.

Kultschitzky's Hematoxylin Staining Solution (Anderson Modification)

Histological Stain

Hematoxylin	1 g
Glacial acetic acid	3 ml
Ethyl alcohol, absolute	10 ml
Calcium hypochlorite solution (2%)	3 ml
Water, distilled	84 ml

Dissolve the hematoxylin in the absolute alcohol. Add the calcium hypochlorite, water, and then the acetic acid. Mix thoroughly and filter.

Leifson's Staining Solution

Flagella Stain

Fuchsin, basic	0.35 g
Sodium chloride	0.50 g
Tannic acid	0.85 g
Ethyl alcohol (95%)	35 ml
Distilled water	65 ml

Mix the fuchsin, sodium chloride, and tannic acid in the water for several minutes and then add the alcohol. Store in a tightly stoppered bottle. Replace the stain after 4 to 5 weeks.

Loeffler's Methylene Blue Staining Solution (Modified)

Bacterial Stain

Methylene blue	0.3 g
Ethyl alcohol (95%)	30 ml
Distilled water	100 ml

Dissolve the methylene blue in the alcohol and then add the water.

Commonly used for bacterial smears.

Loyez Hematoxylin Staining Solution

Histological Stain

Hematoxylin	1 g
Absolute alcohol	10 ml
Lithium carbonate solution (Saturated—1.3 g per 100 ml of water)	2 ml
Distilled water	90 ml

Dissolve the hematoxylin in the alcohol, add the water and then the lithium carbonate solution. Filter and use immediately, since the stain is highly perishable.

Used as a stain for myelin in celloidin and wax sections.

Mann's Staining Solution (Downie's Modification)

Virus Stain

SOLUTION A	
Eosin, yellow	1 g
Orange G	1 g
Distilled Water	100 ml
SOLUTION B	
Methyl blue	1 g
Distilled Water	100 m

The section is first stained with Solution A for 30 minutes, then washed free of stain and stained in Solution B for 5 minutes.

Stain for virus inclusion bodies, which it stains light red, and cell nuclei, blue.

Mayer's Hematoxylin Staining Solution

Histological Stain

Potassium alum	30 g
Hematoxylin	1 g
Sodium iodate (NaIO$_3$)	0.2 g
Water	1000 ml

Heat water to boiling and dissolve the alum in the boiling water. Add the hematoxylin and continue heating until dissolved. Allow to cool and add the sodium iodate. Filter as necessary. The stain needs no ripening and is ready for use. The stain should be prepared fresh every 3 months.

Used for routine staining in botany and zoology. Any counterstain can be subsequently used.

Mayer's Mucicarmine Staining Solution

Stain for Mucin

Carmine	1.0 g
Aluminum chloride, anhydrous	0.5 g
Ethyl alcohol (50%)	100 ml
Distilled water	2 ml

Dissolve the carmine and aluminum chloride in the water. Aid the solution by low heat while stirring constantly until the color turns dark. Gradually add the alcohol with constant stirring and allow to stand for 24 hr. Filter and store in a tightly stoppered bottle.

The mucin stains red.

Mayer's Paracarmine Staining Solution

Protozoan Stain

Carminic acid	1.0 g
Calcium chloride	4.0 g
Aluminum chloride	0.5 g
Ethyl alcohol (70%)	100 ml

Dissolve the ingredients in the alcohol and filter.

Excellent stain for protozoa. The stain may be diluted with additional alcohol (70%) and a drop or two of glacial acetic acid.

Methylene Blue Stain

Stain for Milk Smear

Methylene blue chloride	0.6 g
Ethyl alcohol (95%)	100 ml

Add the methylene blue chloride to the ethyl alcohol and shake for about 5 min. Allow to remain at room temperature for 24 to 48 hr, with occasional shaking, until the dye is completely dissolved. Store in a glass-stoppered bottle.

Meyrick and Harrison Neutral Red Staining Solution

Bacterial Counterstain

Neutral red (1% aqueous solution)	90 ml
Ziehl Nielsen's carbol fuchsin (page 393)	6 ml

Used as a bacterial counterstain. Counterstain for 30 sec only.

Neutral Red Stain

General Stain for Live Organisms

Neutral Red	1.0 g
Water	100 ml

One drop of the dye mixed in 150 ml of water is sufficient to stain an organism red.

Newman-Lampert Stain (Modified by Levowitz and Weber)

Stain for Milk Smear

Methylene blue chloride	0.6 g
Ethyl alcohol (95%)	52 ml
Tetrachlorethane (tech.)	44 ml
Glacial acetic acid	4 ml

Add the methylene blue chloride slowly to the mixture of ethyl alcohol and tetrachlorethane contained in a 200-ml Erlenmeyer flask. Shake the mixture until dissolved and allow to stand for 24 hr at 40°F. Add the glacial acetic acid, filter through a fine textured filter paper. Store the stain in a glass-stoppered bottle.

Nigrosin Staining Solution

Relief Stain

Nigrosin, water-soluble	10 g
Formalin	0.5 ml
Distilled water	90 ml

Boil the nigrosin in the water for about 30 minutes. Filter through filter paper and then add the formalin. Store in small tightly stoppered test tubes.

Not a regular stain. Microorganisms are not stained but appear in relief against a darker background.

North's Methylene Blue and Aniline Oil Stain

Stain for Milk Smear

Methylene blue chloride	0.2 g
Ethyl alcohol (95%)	40 ml
Aniline oil	3 ml
Hydrochloric acid (concentrated)	1.5 ml
Distilled water	55 ml

Dissolve the methylene blue chloride in 30 ml of the ethyl alcohol. Allow to remain at room temperature for 24 to 48 hr with occasional shaking until the dye is completely dissolved.

Mix separately the aniline oil with 10 ml of the ethyl alcohol and then slowly add the hydrochloric acid with constant stirring.

Add the prepared alcoholic methylene blue chloride solution to the aniline oil mixture and then add the distilled water. Filter the solution and store in a glass-stoppered container.

Oil Red O Staining Solution

Stain for Fat

Oil red O	1 g
Alcohol (70%)	50 ml
Acetone	50 ml

If oil red O is not available, use Sudan III or IV.

Used as a special stain for fatty tissue, which it stains red.

Regaud's Hematoxylin Staining Solution

Histological Stain

MORDANT SOLUTION

Iron alum	5 g
Water	100 ml

STAINING SOLUTION

Hematoxylin	1 g
Glycerin	10 ml
Ethyl alcohol (95%)	10 ml
Water	80 ml

DIFFERENTIATING SOLUTION

Picric acid	0.5 g
Ethyl alcohol (95%)	65 ml
Water	35 ml

Ripen the staining solution before use by allowing to stand for a few months.

The specimen is first treated with the mordant solution, washed, transferred to tl staining solution for 30 min at 50°C, washed, and then dipped into the differe tiating solution.

Schiff's Staining Reagent

Stain for Deoxyribonucleic acid

Fuchsin, basic	1 g
Potassium metabisulfite	2 g
Hydrochloric acid, concentrated	2 ml
Charcoal, decolorizing	0.2 g
Distilled water	200 ml

Add the fuchsin to the boiling water and boil until dissolved. Cool the solution to 50°C and add the potassium metabisulfite and stir until dissolved. Cool the solution to room temperature, add the hydrochloric acid and store in the dark for about 24 hr. Add the charcoal, shake the solution for about 5 minutes and filter. Store the solution in the dark. Test the solution with substances containing deoxyribonucleic acid.

Used in the Feulgen reaction for deoxyribonucleic acid. Substances containin deoxyribonucleic acid are colored reddish purple.

Schneider's Carmine Staining Solution

Nuclear Stain

Carmine	0.5 g
Acetic acid (45%)	100 ml

Add the carmine to the boiling acetic acid and allow to dissolve. Cool the solution and filter.

Widely used as a stain for nuclei and chromosomes of fresh cells.

Van Gieson's Staining Solution

Histological Counterstain

Picric acid (saturated—1.2 g per 100 ml distilled water)	10 ml
Acid, fuchsin	0.15 g
Distilled water	65 ml

Bring the ingredients to a boil, cool and filter.

Used as a counterstain for paraffin wax sections.

Weigert's Differentiating Solution

For Staining Procedure

Borax	2 g
Potassium ferricyanide	2.5 g
Distilled water	200 ml

Weigert's Iron Hematoxylin Staining Solution

Histological Stain

STOCK SOLUTION
Dissolve 10 g of hematoxylin in 100 ml of absolute alcohol and allow to ripen for about 5 to 6 weeks by exposing to air and light.

SOLUTION A

Hematoxylin stock solution	10 ml
Ethyl alcohol (95%)	90 ml

SOLUTION B

Ferric chloride solution (50%)	4 ml
Hydrochloric acid, concentrated	1 ml
Distilled water	95 ml

Before using, mix together equal volumes of Solutions A and B and allow to stand about 5 min. A blue-black color is developed. The stain can be kept safely for several days.

Weigert-Van Gieson Staining Solution

Histological Stain

Fuchsin, basic	2 g
Resorcinol	4 g
Ferric chloride (29%)	25 ml
Ethyl alcohol (95%)	200 ml
Hydrochloric acid, concentrated	4 ml
Distilled water	200 ml

Transfer to a porcelain dish the fuchsin, resorcinol and water and boil. Add the ferric chloride to the boiling solution and continue boiling until a precipitate forms (about 5 min). Cool the solution and filter through a paper filter. Transfer the filter paper and precipitate to the porcelain dish and dry thoroughly in an oven at 37°C. Add the alcohol to the dry powder and boil carefully on a hot plate, stirring constantly, until the powder is completely dissolved. Cool and filter. Add the hydrochloric acid to the staining solution.

Used as an identifying stain for connective and elastic tissue.

Wolbach's Giemsa Staining Solution

Histological Stain

Giemsa's stock solution (see page 381)	1.25 ml
Methyl alcohol	1.50 ml
Distilled water	50 ml

An excellent stain for sections and for films and smears. Mount in cedarwood oil or clarite, as the stain will fade with a balsam mount.

Wright's Stain

Blood Film Stain

WRIGHT'S STAIN

Wright's stain powder	0.3 g
Glycerin	3.0 ml
Methyl alcohol (absolute)	97.0 ml

Grind the Wright's stain powder and the glycerin together in a mortar. The Wright's stain powder is a specially prepared methylene blue-eosine mixture. Add the methyl alcohol and mix. Transfer the solution to a flask, stopper tightly and allow to stand for 24 hr.   Filter the solution and again allow to stand several days before using.   If the humidity is high, the glycerin may be omitted.

BUFFER SOLUTION

Potassium acid phosphate	1.63 g
Disodium phosphate	3.20 g
Distilled water	1000 ml

Wright's Stain

Alternate Formula

Wright's stain powder	1.5 g
Disodium phosphate	0.6 g
Potassium acid phosphate	0.4 g
Methyl alcohol (absolute)	500 ml

Grind the solid ingredients together in a mortar and then add the alcohol. Allow to stand for several days, stirring occasionally.   Filter the solution prior to use.

If the alternate formula is used, add distilled water instead of the buffer solution during the staining procedure.

Ziehl-Neelsen's Carbol Fuchsin Staining Solution

Bacterial Stain

Fuchsin, basic (10% alcoholic solution)	10 ml
Phenol (5% aqueous solution)	100 ml

Mix the ingredients and filter.

Used for the detection and differentiation of acid-fast bacilli.

MISCELLANEOUS HISTOLOGICAL REAGENTS

Acidified Alcohol Solution

For Differentiation of Stains

Hydrochloric acid, concentrated	1 ml
Ethyl alcohol (70%)	99 ml

Anticoagulant Solution

Blood Coagulation Preventive

Ammonium oxalate	1.2 g
Potassium oxalate	0.8 g
Distilled water	100 ml

Dissolve the oxalates in the water. Solution is used in proportion of 0.1 ml to 1 ml of blood.

Berlese's Medium

Gum Mountant

Gum acacia	8 g
Chloral hydrate	75 g
Dextrose syrup	5 ml
Glacial acetic acid	3 ml
Water	10 ml

Mix together the water, acetic acid, and dextrose syrup; dissolve the gum acacia in this slution. It may take a week or more for the gum to dissolve. Stir the mixture from time to time so that few air bubbles are included in the mixture.

When solution of the gum is complete, add the chloral hydrate and stir.

A medium with a high index of refraction, used for specimens where high transparency equivalent to balsam mounts is required. If quicker drying is required, add 60 rather than 75 g of chloral hydrate.

Canada Balsam (Acid)

Mountant

Add sufficient *solid Canada balsam* to *xylene* so that a syrupy consistency will be achieved after the balsam is dissolved. Aid the solution of the balsam by placing in an incubator maintained at 37°C. Add a pinch of *salicylic acid* and then mix gently by stirring. Allow the solution to settle, then decant the clear xylene-balsam mixture.

Canada Balsam (Neutral)

Mountant

Add sufficient *solid Canada balsam* to *xylene* so that a syrupy consistency will be achieved after the balsam is dissolved. Aid the solution of the balsam by placing in an incubator maintained at 37°C.   Add a pinch of *calcium carbonate* and then gently mix by stirring.   Allow the solution to settle, then decant the clear xylene-balsam mixture.

Chromic Acid Cleaning Solution

Glass Cleaner

Chromium trioxide (saturated—about 170 g per 100 ml of water)	25 ml
Sulfuric acid, technical	2220 ml
	(9 lb)

Add the sulfuric acid very carefully to the chromium trioxide.   A rise in temperature of about 15°C will occur.   Some of the chromium trioxide will separate as scarlet crystals and form a reservoir of dichromate ions. *Extreme precaution* must be observed in handling sulfuric acid.

Egg Albumin Substitute

Adhesive for Glass Slides

Water glass	33 ml
Distillled water	100 ml

Farrant's medium
Gum Mountant

Gum acacia	40 g
Glycerin	20 ml
Phenol	0.1 g
Water	40 ml

A medium with a low index of refraction; used for specimens where high transparency is not required.   When applied to a slide, it requires several days for drying.   Place the gum acacia in a piece of muslin and suspend in the water until the gum is in solution.   Remove the muslin from the solution and add the glycerin and phenol.   If necessary, filter the mixture through wet lint.

Glycerin Gel

Mountant

Gelatin	3 g
Glycerin	20 ml
Chromium potassium sulfate	0.2 g
Distilled water	80 ml

Dissolve the gelatin in 50 ml of water. Aid the solution by means of heat.

Dissolve separately the chromium potassium sulfate in 30 ml of the water. Aid the solution by means of heat.

Add the glycerin to the warm gelatin solution and then add the warm chrome alum solution while constantly stirring. Filter and add a crystal of thymol as a preservative. If bubbles are present transfer to an oven maintained at 37°C until the air bubbles disappear.

Used when the specimen is mounted directly from water by melting a small portion of the gel on the slide.

Gray and Wess's Medium

Gum Mountant

Polyvinyl alcohol	2 g
Glycerin	5 ml
Acetone (70%)	7 ml
Lactic acid	5 ml
Water	10 ml

Add the polyvinyl alcohol to the acetone and mix until a smooth paste is achieved. Mix together 5 ml of water, glycerin, and lactic acid and add the mixture to the paste while stirring. Add the remaining 5 ml of water, drop by drop, to the mixture, stirring constantly. The mixture, which is cloudy, is heated on a water bath for about 10 min, or until transparent.

A medium with a low index of refraction; used for specimens where high transparency is not required. When applied to a slide, it sets rapidly and can be handled safely after half an hour.

Hamilton's Freezing Mixture

For Preparation of Frozen Sections

SOLUTION A

Cane sugar	142.5 g
Distilled water	150 ml
Formaldehyde (40%)	9 ml

Dissolve the cane sugar in the water with the aid of heat. Allow to cool and filter through muslin. Add the formaldehyde to prevent the growth of molds.

SOLUTION B

Gum acacia	2.8 g
Distilled water	150 ml
Formaldehyde	5 ml

Place the gum acacia in a piece of muslin and suspend in water until the gum is in solution. Filter through muslin and add the formaldehyde to prevent the growth of molds.

For use mix equal parts of Solutions A and B.

A thin layer of Hamilton's freezing mixture is placed over the surface of the freezing stage just before a frozen section is prepared.

Hanley's Solution

Narcotic for Microscopic Organisms

Eucaine hydrochloride	0.3 g
Ethyl cellosolve	10 ml
Water	90 ml

An excellent narcotic for small organisms such as the rotifers. For use, add 1 drop to 10 ml of water.

Jurray's Mixture

For Softening Chitinous Materials

Phenol	25 g
Chloral hydrate	25 g

The chitinous organism, after fixing, is transferred to this mixture for 24 hr. The mixture is then washed free with chloroform.

Lenoir's Fluid

For Removal of Picric Acid After Fixation

Ammonium acetate	10 g
Ethyl alcohol (95%)	30 ml
Water	70 ml

Fluid for liberating the picric acid bound to the specimen during fixation (picric acid may interfere with staining).

Lugol's Iodine Solution

For Removal of Mercury Salts and Picric Acid After Fixation

Potassium iodide	1 g
Iodine crystals	0.5 g
Water	50 ml

To 3 ml of water, add the potassium iodide and the iodine. Stir until dissolved. Add the remainder of the water.

Also used as a fixing agent for small marine animals and spermatozoa.

Mayer's Egg Albumin Mixture

Adhesive for Glass Slides

Egg white	50 ml
Glycerin	50 ml
Sodium salicylate	1 g

Mix the ingredients thoroughly and filter through a wet cloth.

Miquel Solution

For Cultivation of Diatoms

SOLUTION A

Potassium nitrate	20.2 g
Distilled water	100 ml

SOLUTION B

Calcium chloride ($CaCl_2.6H_2O$)	4 g
Disodium phosphate ($Na_2HPO_4.12H_2O$)	4 g
Ferric chloride ($FeCl_3.6H_2O$) (melted)	2 ml
Hydrochloric acid, concentrated	2 ml
Distilled water	80 ml

To prepare Solution B, dissolve the calcium chloride in 40 ml of distilled water and add the hydrochloric acid. In a separate beaker, dissolve the disodium phosphate in 40 ml of distilled water, add the melted ferric chloride and slowly mix these two solutions.

To prepare Miquel's solution, add 2 ml of Solution A and 1 ml of Solution B to 1 liter of sea water. Sterilize this solution by bringing just to the boiling point. Cool the solution, decant or filter off the slight precipitate. Divide the remaining solution into two 1 liter flasks. Shake vigorously to aerate the solution.

Pour the Miquel's solution into sterile, wide-mouth, 125-ml flasks, to half their capacities, so that a large surface of the solution is in contact with the air. Inoculate the flasks by adding 6 to 8 ml of an old culture of diatoms. Place them near a window but not in direct sunlight. Shake the flasks at least once a day. The cultures grow best at about 15 to 16°C.

Priman's Egg Albumin Substitute

Adhesive for Glass Slides

Serum, human, fresh	15 ml
Formaldehyde (5%)	6 ml
Distilled water	10 ml

Mix the ingredients and filter through a filter paper.

Ringer's Solution (Normal Saline)

For Culture Work

Sodium chloride	8.0 g
Calcium chloride	0.2 g
Potassium chloride	0.2 g
Magnesium chloride	0.1 g
Monosodium phosphate	0.1 g
Sodium bicarbonate	0.4 g
Distilled water	1,000 ml

The monosodium phosphate and sodium bicarbonate are dissolved separately in a small volume of sterile distilled water, filtered through an asbestos filter, and then added to the sterile solution. The remainder of the ingredients are dissolved in the water and autoclaved at 15 lb pressure for 20 min.

The solution approximates the blood serum of animals.

Scott's Tap Water Substitute

Used for Blueing

Sodium bicarbonate	3.5 g
Magnesium sulfate	20 g
Tap water	1000 ml

To prevent the growth of molds, add a 1% formalin solution

Schreiber's Solution

For Cultivation of Diatoms and Green Forms

Potassium nitrate (KNO_3)	0.2 g
Dipotassium hydrogen phosphate (K_2HPO_4)	0.1 g
Potassium silicate (K_2SiO_3)	0.01 g
Ferric sulfate ($Fe_2(SO_4)_3$)	0.005 g
Redistilled water	50 ml

This solution is added to 950 ml of filtered seawater and the medium is sterilized by steam at 100°C. If a precipitate (calcium carbonate) is formed during the heating, filter the solution before use.

Starch Paste

For Adhesive Glass Slides

Corn starch, powdered	1 g
Hydrochloric acid (10%)	2 drops
Water	30 ml
Thymol	small crystal

Add the starch to 10 ml of cold water and mix thoroughly. Pour the mixture into 20 ml of boiling water. Add the 2 drops of the hydrochloric acid and boil for 5 min while stirring constantly. Allow to cool and add a small crystal of thymol.

Used as a substitute for Mayer's egg albumin mixture.

PHOTOGRAPHIC SOLUTIONS

Metol-Hydroquinone Contrast Developing Solution

Photographic Developer

Metol or elon	1.0 g
Sodium sulfite, desiccated	75.0 g
Hydroquinone	9.0 g
Sodium carbonate, desiccated	25.0 g
Potassium bromide	5.0 g
Water, sufficient to make	1 liter

Dissolve the chemicals, in the order given, in 500 ml of water maintained at about 50°C. Add additional water to make 1 liter of developer.

Solution for developing plates and films with maximum contrast. Average developing time, in a tank and without dilution, is about 5 min at 20°C. For tray development, 3 to 4 min with constant agitation is satisfactory. If less contrast is desired, dilute the solution with water, 1:1.

Metol-Hydroquinone Developing Solution (D-55)

Photographic Print Developer

Metol	2.4 g
Sodium sulfite, desiccated	36.0 g
Hydroquinone	10.0 g
Sodium carbonate, desiccated	36.0 g
Potassium bromide	12.0 g
Water	1 liter

Dissolve the chemicals in the order given in 1 liter of water.

Solution recommended for developing bromide enlarging papers. For use mix 1 vol of the solution with 2 vol of water. Develop prints from 2 to 3 min.

Stop Bath for Prints and Films

Photographic Short-Stop

Acetic acid (28%)	50 ml
Water	1000 ml

Leave prints or film in the solution for a minimum of 5 sec.

Stop Bath for Films

Photographic Hardener

Sulfuric acid (5%)	30 ml
Potassium chrome alum	30 g
Water, sufficient to make	1 liter

Agitate the solution for a few seconds after immersing the negative. Leave the negative in the bath for 3 to 5 min.

Metol-Hydroquinone Developing Solution (D-72)

Photographic Developer

Metol or elon	3 g
Sodium sulfite, desiccated	45 g
Hydroquinone	12 g
Sodium carbonate, monohydrate	80 g
Potassium bromide	2 g
Water, sufficient to make	1 liter

Dissolve the chemicals in the order given in 500 ml of water maintained at about 50°C. Add additional water to make 1 liter.

Solution is a good general developer for films, plates, slides, and printing papers. The average developing time for orthochromatic films or plates, to achieve average contrast, is 4 to 5 min at 20°C when 1 vol of the solution is diluted with 1 vol of water. If maximum density is required, the solution is used full strength. For paper prints, dilute 1 vol of solution with 1 vol of water to achieve cold tones; for warm tones, dilute 1 vol of solution with 2 vol of water.

Metol-Hydroquinone Developing Solution (D-76)

Photographic Developer

Metol or elon	2 g
Sodium sulfite, desiccated	100 g
Hydroquinone	5 g
Borax, granular	2 g
Water, sufficient to make	1 liter

Dissolve the chemicals in the order given in 750 ml of water maintained at about 50°C. Add additional water to make 1 liter.

Solution for developing plates and films with normal contrast, fine grain, and at good speed. The average developing time, when using the developer without dilution in a tank, is about 12 min at 20°C. For tray development, 7 to 10 min with constant agitation is satisfactory.

Intensifying Solution

Photographic Intensifier

Mercuric iodide	2 g
Potassium iodide	2 g
Potassium thiosulfate	2 g
Water	100 ml

Wash the negatives well before placing in the solution. When the desired density has been reached, remove the negative and rinse thoroughly in water.

To make the intensification of the negative permanent, immerse it in a 1% sodium sulfide solution until the image assumes a brownblack color.

Ammonium Persulfate Reducing Solution (R-1)

Photographic Reducer

Ammonium persulfate	56 g
Sulfuric acid (10%)	30 ml
Water	970 ml

For use, dilute 1 volume of the solution with 2 volumes of water. This solution minimizes the contrast of the negative as it reduces.

After reduction, place the negative in a solution of sodium sulfite to neutralize any persulfate remaining in the negative emulsion and simultaneously stop the reducing action.

Farmer's Reducing Solution

Photographic Reducer

SOLUTION A

Potassium ferricyanide	40 g
Water	500 ml

SOLUTION B

Sodium thiosulfate	230 g
Water	1 liter

Just before use, mix 1 vol of Solution B with 7 vol of water. Add a little of Solution A to this mixture so that it is colored yellow.

The solution increases the contrast as it reduces the density of the negative.

Permanganate Reducing Solution (R-2)

Photographic Reducer

SOLUTION A

Potassium permanganate	49 g
Water	1 liter

SOLTION B

Sulfuric acid solution (6%)	1 liter

Just prior to use, mix 1 vol of Soution A, 2 vol of Solution B, and 64 vol of water.

The solution increases the contrast as it reduces the density of the negative. After reduction, place the negative in hypo for a few minutes to remove any yellow stain and then wash thoroughly.   The solution aids in the removal of any brown developer stains.

Acid Hardening Fixing Solution

Photographic Fixer

SOLUTION A (HYPO)

Sodium thiosulfate	285 g
Water, sufficient to make	1 liter

SOLUTION B

Sodium sulfate, desiccated	75 g
Acetic acid (28%)	235 ml
Boric acid, crystals	37.5 g
Potassium alum	75 g
Water, sufficient to make	1 liter

Dissolve the chemicals in Solution B in 600 ml of water maintained at 50°C.   Allow to cool and dilute to 1 liter.   Before use, mix 4 vol of Solution A with 1 vol of Solution B.

An acid hardening-fixing solution that hardens the emulsion and permits a higher drying temperature.   The solution is suitable for negatives and prints.

Hypo Test Solution

For Testing the Removal of Hypo from Films and Plates

Silver nitrate	7.5 g
Acetic acid (28%)	125 ml
Water	875 ml

Dissolve the silver nitrate in the acetic acid solution and water.   Store the solution in an amber, tightly stoppered bottle away from light.

To determine whether the washing has been thorough, cut a very small blank strip from the washed print or film and immerse it in a portion of the solution for about 3 min.   Any strong discoloration in the film or paper indicates that it has not been washed sufficiently.   A very light discoloration may be disregarded.

STAIN FOR CHEMICAL MICROSCOPY

Hertzberg's Stain

Fiber Stain

SOLUTION A

Dissolve 50 g of zinc chloride in 25 ml of distilled water.

SOLUTION B

To 12.5 ml of distilled water, add 5.25 g of potassium iodide and 0.25 g of iodine.

Add Solution A to Solution B and mix. Allow to stand for 24 hr in a dark place. Decant the clear liquid into a small amber bottle, add a crystal of iodine, and store in a dark place. If the stain imparts a light-blue or purplish color to sulfate fibers, add more zinc chloride to intensify the blue color. The zinc chloride imparts the blue color to chemical wood pulp while iodine produces the wine color in rag fibers and yellow color in ground-wood. Prepare a fresh solution every few months.

Used in the identification and quantitative estimation of the types of fibers present in paper and some textiles of vegetable origin. The stain gives definite identification and quantitative estimation of sulfite (blue), groundwood (yellow), and rag (wine or purplish pink) fibers in paper.

INDEX

www.ingramcontent.com/pod-product-compliance
Lightning Source LLC
Chambersburg PA
CBHW021428180326
41458CB00001B/178